UMAP

Modules

Tools for Teaching 1999

published by

The Consortium for Mathematics
and Its Applications, Inc.
Suite 210
57 Bedford St.
Lexington, MA 02420

edited by

Paul J. Campbell
Campus Box 194
Beloit College
700 College St.
Beloit, WI 53511–5595
campbell@beloit.edu

ISBN 0–912843–70–5
Typeset and printed in the U.S.A.

Subscription Rates for 2000 Calendar Year: Volume 21

The UMAP Journal is published quarterly by the Consortium for Mathematics and Its Applications (COMAP), Inc., Suite 210, 57 Bedford Street, Lexington, MA, 02420, in cooperation with the American Mathematical Association of Two-Year Colleges (AMATYC), the Mathematical Association of America (MAA), the National Council of Teachers of Mathematics (NCTM), the American Statistical Association (ASA), the Society for Industrial and Applied Mathematics (SIAM), and The Institute for Operations Research and the Management Sciences (INFORMS). The Journal acquaints readers with a wide variety of professional applications of the mathematical sciences and provides a forum for the discussion of new directions in mathematical education (ISSN 0197-3622).

Table of Contents

Introduction

The instructional Modules in this volume were developed by the Undergraduate Mathematics and Its Applications (UMAP) Project. Project UMAP develops and disseminates instructional modules and expository monographs in mathematical modeling and applications of the mathematical sciences, for undergraduate students and their instructors.

UMAP Modules are self-contained (except for stated prerequisites) lesson-length instructional units. From them, undergraduate students learn professional applications of the mathematical sciences. UMAP Modules feature different levels of mathematics, as well as various fields of application, including biostatistics, economics, government, earth science, computer science, and psychology. The Modules are written and reviewed by instructors in colleges and high schools throughout the United States and abroad, as well as by professionals in applied fields.

UMAP was originally funded by grants from the National Science Foundation to the Education Development Center, Inc. (1976–1983) and to the Consortium for Mathematics and Its Applications (COMAP) (1983–1985). In order to capture the momentum and success beyond the period of federal funding, we established COMAP as a nonprofit educational organization. COMAP is committed to the improvement of mathematics education, to the continuing development and dissemination of instructional materials, and to fostering and enlarging the network of people involved in the development and use of materials. In addition to involvement at the college level through UMAP, COMAP is engaged in science and mathematics education in elementary and secondary schools, teacher training, continuing education, and industrial and government training programs.

In addition to this annual collection of UMAP Modules, other college-level materials distributed by COMAP include individual Modules (more than 500), *The UMAP Journal*, and UMAP expository monographs. Thousands of instructors and students have shared their reactions to the use of these instructional materials in the classroom, and comments and suggestions for changes are incorporated as part of the development and improvement of materials.

This collection of Modules represents the spirit and ability of scores of volunteer authors, reviewers, and field-testers (both instructors and students). The substance and momentum of the UMAP Project comes from the thousands of individuals involved in the development and use of UMA instructional materials. COMAP is very interested in receiving information on the use of Modules in various settings. We invite you to call or write for a catalog of available materials, and to contact us with your ideas and reactions.

Sol Garfunkel, COMAP Director
Paul J. Campbell, Editor

Recruiting, editing, and selecting UMAP Modules is done by the board of editorial board of *The UMAP Journal*, who are appointed by the editor-in-chief in consultation with the presidents of the cooperating organizations:

Mathematical Association of America (MAA),
Society for Industrial and Applied Mathematics (SIAM),
National Council of Teachers of Mathematics (NCTM),
American Mathematical Association of Two-Year Colleges (AMATYC),
Institute for Operations Research and the Management Sciences
(INFORMS), and
American Statistical Association (ASA).

In 1999, the editorial board members were:

Manuscripts are read double-blind by two or more referees, including an associate editor. Guidelines for authors are published in the first issue of each volume of *The UMAP Journal*; COMAP's copyright policy and copyright release form appear in 18 (1997)(1): 1–14. The guidelines and copyright material are available also at `ftp://cs.beloit.edu/math-cs/Faculty/Paul Campbell/Public/UMAP` .

UMAP

Modules in
Undergraduate
Mathematics
and Its
Applications

Published in
cooperation with

The Society for
Industrial and
Applied Mathematics,

The Mathematical
Association of America,

The National Council
of Teachers of
Mathematics,

The American
Mathematical
Association of
Two-Year Colleges,

The Institute for
Operations Research
and the Management
Sciences, and

The American
Statistical Association.

Module 775

The Resilience of Grassland Ecosystems

Ray Huffaker
Kevin Cooper
Thomas Lofaro

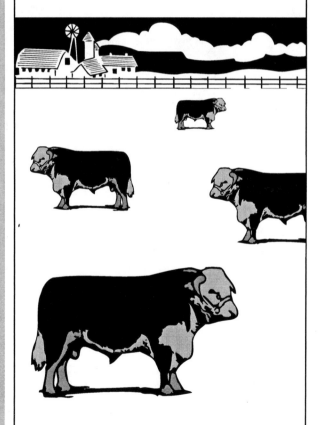

**Applications of Differential Equations
to Biology and Ecology**

INTERMODULAR DESCRIPTION SHEET:	UMAP Unit 775
TITLE:	The Resilience of Grassland Ecosystems

AUTHORS:

Kevin Cooper
Department of Pure and Applied Mathematics
kcooper@pi.math.wsu.edu

Ray Huffaker
Department of Agricultural Economics

Thomas Lofaro
Department of Pure and Applied Mathematics

Washington State University
Pullman, WA 99164

MATHEMATICAL FIELD:	Differential equations
APPLICATION FIELD:	Biology, ecology
TARGET AUDIENCE:	Students in a course in differential equations

ABSTRACT:
This Module introduces students to the *state-and-transition theory* explaining the succession of plant species on grassland and to the concept of *successional thresholds* partitioning plant states into those gravitating toward socially desirable or socially undesirable plant compositions over time. Students are shown how the state-and-transition theory is formulated in the mathematical ecology literature as a system of two autonomous differential equations, and how a successional threshold is defined by the stable manifold to an interior saddle-point equilibrium. A series of exercises directs students toward a qualitative phase-plane solution of the system and an analytical approximation of the stable manifold. Students also gain experience working with the numerical phase-plane plotter Dynasys, which can be downloaded from the World Wide Web. A discussion section applies the approximated stable manifold to the real-world problem of controlling livestock numbers on public grazing land to reestablish more socially desirable plant varieties. The Module is within the capabilities of students having had basic calculus and an introductory course in ordinary differential equations covering phase-plane solutions.

PREREQUISITES:	Introduction to ordinary differential equations covering phase-plane solutions.

Tools for Teaching 1999, 1–17. Reprinted from *The UMAP Journal* 20 (1) (1999) 29–45.
©Copyright 1999, 2000 by COMAP, Inc. All rights reserved.

COMAP, Inc., Suite 210, 57 Bedford Street, Lexington, MA 02420
(800) 77-COMAP = (800) 772-6627, or (781) 862-7878; http://www.comap.com

The Resilience of Grassland Ecosystems

Kevin Cooper
Department of Pure and Applied Mathematics
kcooper@pi.math.wsu.edu

Ray Huffaker
Department of Agricultural Economics

Thomas Lofaro
Department of Pure and Applied Mathematics
Washington State University
Pullman, WA 99164

Table of Contents

MODULES AND MONOGRAPHS IN UNDERGRADUATE
MATHEMATICS AND ITS APPLICATIONS (UMAP) PROJECT

The goal of UMAP is to develop, through a community of users and developers, a system of instructional modules in undergraduate mathematics and its applications, to be used to supplement existing courses and from which complete courses may eventually be built.

The Project was guided by a National Advisory Board of mathematicians, scientists, and educators. UMAP was funded by a grant from the National Science Foundation and now is supported by the Consortium for Mathematics and Its Applications (COMAP), Inc., a nonprofit corporation engaged in research and development in mathematics education.

Paul J. Campbell   Editor
Solomon Garfunkel   Executive Director, COMAP

1. Introduction

Grass varieties compete with one another for habitat in grassland ecosystems. Native grasslands in the intermountain region of the United States are dominated by highly competitive perennial grasses (e.g., bluestem, grama, and bunch grasses) as understory species to sagebrush. However, historic overgrazing of these native grasses by domestic livestock has reduced the grasses' vigor, and consequently their ability to withstand the invasion of highly competitive alien annual grasses introduced inadvertently by settlers, principally cheatgrass (*Bromus tectorum L.*) [Evans and Young 1972]. Cheatgrass has competed so strongly that it currently dominates millions of acres of native grasslands in the intermountain West [Evans and Young 1972]. It is not valueless in livestock production, but livestock do significantly less well on it than they do on native species. Cheatgrass also promotes several environmental problems. It is more superficially rooted than the relatively large fibrous root systems of perennials, and thus is not well suited for binding soil. This promotes soil erosion that, among other problems, harms riparian habitat for fish and wildlife [Stewart and Hull 1949].

The competitive dominance of cheatgrass in the intermountain West, and other alien grasses in other regions of the world, has led grassland ecologists to question the extent to which the underlying competitive forces can be reversed so that more beneficial native grasses again dominate. The conventional *plant-succession theory* contends that the variety of possible plant compositions of a grassland ecosystem is a hierarchy of *successional states*. An intervening factor such as livestock grazing can cause a retrogression in successional states from a *climax state* including only native plant varieties to lower successional states including less desirable alien varieties. When the intervening factor is removed, the grassland ecosystem is claimed to undergo a *secondary succession*, wherein the system reverses along the same pathway of successional states toward the stable climax state.

The plant-succession theory has begun to lose ground to the recently introduced *state-and-transition theory* due to empirical evidence that grassland ecosystems are not so resilient after the removal of an intervening factor. The state-and-transition theory predicts less optimistically that an intervening factor may result in plant compositions that are locked into *basins of attraction* compelling them toward stable lower successional states over time [Westoby et al. 1989; Laycock 1991]. The state-and-transition theory implies that the success of livestock reductions in promoting secondary succession depends on whether grassland conditions can be pushed across thresholds of environmental change to more socially desirable stable plant states. Thus, successional thresholds become the key analytical tool in characterizing the resilience of grassland ecosystems.

Boyd [1991] formulated the state-and-transition theory as a special case of Gause's interspecies competition equations (see. e.g., Hastings [1996]) to study, among other things, how selective grazing by wildlife on perennial grasses is linked to the long-term successional change in the plant composition of grassland. This Module contains a series of exercises deriving the necessary and sufficient ecological conditions under which Boyd's formulation results in a stable manifold partitioning phase space into basins of attraction to equilibria that represent desirable and less-desirable plant compositions. The stable manifold is composed of the convergent separatrices associated with an interior saddle-point equilibrium and represents the successional threshold of the grassland ecosystem in recovering from historic overgrazing. Further exercises demonstrate how the successional threshold can be approximated analytically using the theory of eigenvalues and eigenvectors and how the approximation can be analyzed for its mathematical accuracy.

2.   Mathematical Formulation of the State-and-Transition Theory

Let Ω denote an area in which perennial and annual grasses are in competition. Let the variable $0 < g < 1$ denote the fraction of Ω colonized by grasses (i.e., perennial grasses) and the variable $0 < w < 1$ denote the fraction colonized by weeds (i.e., annual grasses). Portions of Ω may be barren or have overlapping vegetation, thus the sum of g and w need not equal one. The equations describing their competitive dynamics through time (t) are

$$g' = \{r_g [1 - g - q_w(g, E)w]\} g \tag{1a}$$
$$w' = \{r_w [1 - g - q_g(g, E)g]\} w. \tag{1b}$$

The bracketed terms multiplying g on the right-hand side (RHS) of **(1a)** and w on the RHS of **(1b)** measure the net per capita colonization rates of grasses and weeds, respectively. Parameters r_g and r_w (both with dimension $1/t$) measure the intrinsic colonization rates of g and w as each approaches zero, and $q_g(g, E)$ and $q_w(g, E)$ are nonnegative unitless competition rates that vary with g and a parameter E. The competition rates are discussed below.

Assume momentarily that q_g and q_w are zero (i.e., that grasses and weeds are not competitive). Equations **(1a)** and **(1b)** collapse to

$$g' = r_g g(1 - g) \tag{2a}$$
$$w' = r_w w(1 - w), \tag{2b}$$

which are basic logistic growth functions.

Exercise

1. Equations **(2a)** and **(2b)** are essentially alike, so we can examine the behavior of just one of them. Graph **(2b)** and use the graph to determine equilibrium colonization levels (i.e., values w^e for which w' equals zero). Solve the equation and graph its solution over $0 < w < 1$. What is the behavior of the solution around the equilibrium colonization levels w^e?

Positive competition rates q_g and q_w in **(1a)** and **(1b)** relate the colonization rate of one plant group to the other's competitive loss of habitat and have the impact of decreasing the loser's net per capita colonization rate. Each competition rate is a function of g because the density of grasses is assumed to determine its ability to compete with weeds for habitat. Grasses compete more favorably when g increases, forcing q_g toward an upper bound q_g^u and q_w toward a lower bound q_w^l. Boyd [1991] models the direct (inverse) bounded relationship between g and q_g (q_w) with the following *Michaelis-Menten functions*:

$$q_g = q_g^u \left(\frac{BE + g}{E + g} \right) \tag{3a}$$

$$q_w = q_w^l \left(\frac{E + g}{BE + g} \right). \tag{3b}$$

The parameter B is the ratio between the lower and upper bounds on q_g and q_w, that is, $B = q_g^l / q_g^u = q_w^l / q_w^u$. To simplify the model, Boyd assumes that B is the same for both grasses and weeds. As g increases, **(3a)** increases q_g from its lower bound ($q_g^l = Bq_g^u$ when $g = 0$) asymptotically toward its upper bound (q_g^u as $g \to \infty$). Conversely, as g increases, **(3b)** decreases q_w from its upper bound ($q_w^u = q_w^l / B$ when $g = 0$) asymptotically toward its lower bound (q_w^l as $g \to \infty$). The competition rates respond more rapidly to increases in g (i.e., q_g and q_w approach their maximum and minimum values, respectively, at lower levels of g) for small levels of the parameter E.

Exercise

2. Graph **(3a) and (3b)** to verify the above properties.

Because the competitiveness of grasses is inversely related to E, we can account for the adverse impact of livestock grazing on the competitiveness of grasses indirectly by fixing E at a relatively high level when the number of livestock preferentially grazing grasses in area Ω is relatively large. Alternatively, we can fix E at a relatively low level when the grazing pressure exerted on Ω is relatively small.

3. Solution Analysis

Inserting **(3a)** and **(3b)** into **(1a)** and **(1b)** yields:

$$g' = r_g g \left(1 - g - q_w^l \frac{E + g}{BE + g} w \right) \tag{4a}$$

$$g' = r_w w \left(1 - w - q_g^u \frac{BE + g}{E + g} g \right). \tag{4b}$$

We will solve this system using conventional phase diagram techniques, and begin by deriving the *nullcline functions* found by setting $g' = w' = 0$ in (g, w)-space.

Exercises

3. Each of the equations **(4a)** and **(4b)** yields a pair of nullclines. One of the nullclines from setting $g' = 0$ is the w-axis and one of the nullclines from setting $w' = 0$ is the g-axis. Analytically solve for the other two interior nullclines. Denote the interior nullcline from setting $g' = 0$ as $N_g(g)$, and that from setting $w' = 0$ as $N_w(g)$.

4. Verify that $N_g(g)$ has a g-axis intercept at 1, where grasses are 100% colonized and weeds are extinct, and a w-axis intercept at critical level $w_c = 1/q_w^u$. Also verify that $N_w(g)$ has a w-axis intercept at 1, where weeds are 100% colonized and grasses are extinct, and a g-axis intercept at critical level

$$g_c = \frac{1 - EBq_g^u + \sqrt{\left(EBq_g^u - 1\right)^2 + 4Eq_g^u}}{2q_g^u}. \tag{5}$$

Use **(5)** to solve for the critical value of parameter $E = E_c$ that sets $g_c = 1$. Observe that g_c is less (greater) than one when E is set at a value less (greater) than E_c.

The number of steady-state solutions and their stability properties turn out to depend largely on the relative magnitudes of critical levels w_c and g_c to one. We will study the configurations leading to the existence of thresholds partitioning phase space into basins of attraction gravitating toward equilibrium plant states of differing desirability.

Exercise

5. Assume the following parameter values:
$$r_g = 0.27, \quad r_w = 0.35, \quad q_w^l = 0.6, \quad q_g^u = 1.07, \quad \text{and} \quad B = 0.3.$$
Plot $N_g(g)$ and $N_w(g)$ on the same graph for each of the following three cases:

a) $E = 0.4$;

b) $E = 0.172$; and

c) $E = 0.06$.

Recall that the parameter E is inversely related to the competitiveness of grasses, so that the impact of decreasing E over the three cases is to make grasses increasingly competitive. Identify the steady-state solutions on the axes and in the interior of the three plots. Use **(4a)** and **(4b)** to determine the directions of motion of w and g over time in the various areas partitioned by the nullclines and draw in the requisite trajectories. Comment on the observed stability of each steady-state solution.

Your work in **Exercise 5** should show that the phase-diagram solution for each of the three cases has an unstable node equilibrium at the origin ($w = g = 0$), an all-weeds equilibrium along the w-axis ($g = 0$, $w = 1$), and an all-grasses equilibrium along the g-axis ($g = 1$, $w = 0$). Stability of the all-weeds equilibrium depends completely on critical level w_c (w-axis intercept of the grasses nullcline N_g), which is inversely related to the upper bound on the competitive ability of weeds, that is, $w_c = 1/q_w^u$. When $w_c < 1$ ($q_w^u > 1$), as in all three cases above, weeds are a relatively strong competitor and the all-weeds equilibrium is a stable node attracting all initial plant states, including some level of grasses. Stability of the all-grasses equilibrium depends completely on critical level g_c (g-axis intercept of the weeds nullcline N_w, equation **(5)**), which in turn depends on the magnitude of the parameter E with respect to critical level E_c (see **Exercise 4**). For the parameter values in **Exercise 5**, $E_c = 0.103093$. When $g_c > 1$ ($E > E_c$), as in cases **(a)** and **(b)** in **Exercise 5**, grasses are a relatively weak competitor and the all-grasses equilibrium is a saddle point, repelling all plant states including some level of weeds in the positive quadrant. Alternatively, when $g_c < 1$ ($E < E_c$), as in case **c)** in **Exercise 5**, grasses are a relatively strong competitor and the all-grasses equilibrium is a stable node attracting a range of initial plant states.

Exercise

6. Use a computer program to generate numerically the phase diagrams associated with the three cases in **Exercise 5**. One such program that you can download from the World Wide Web is DynaSys at http://www.sci.wsu.edu/idea/software.html .

4.   Successional Thresholds

Cases **(b)** and **(c)** in **Exercise 5** generate an interior saddle-point equilibrium sandwiched between two exterior stable nodes. One exterior stable node is always the all-weeds equilibrium. The other exterior stable node is the all-grasses equilibrium when $N_g(g)$ and $N_w(g)$ intersect once in the positive quadrant (case **(c)**), or an equilibrium with some level of weeds when $N_g(g)$ and $N_w(g)$ intersect twice in the positive quadrant (case **(b)**).

Exercise

7. Using the parameter values from **Exercise 5** with E fixed at 0.172, choose a set of initial conditions resting on the line $w = 0.2$ in the phase plane. Use a computer program to plot the phase trajectories passing through these initial conditions. What happens to the trajectories as the initial conditions move to the left? Try to find a curve that divides the set of initial conditions on trajectories approaching the undesirable all-weeds equilibrium at $(0, 1)$ from those on trajectories approaching an equilibrium with some level of grasses. How did you get the program to draw that curve?

The curve that separates the regions of different behavior in the phase plane is called a *separatrix*. One separatrix joining the interior saddle point emanates upward from the unstable node at the origin. Another separatrix converges downward to join the interior saddle point. Both separatrices taken together comprise the *stable manifold* of the saddle-point equilibrium [Hale and Koçak 1991]. The stable manifold partitions plant states into two disjoint groups. All plant states to the left of the manifold gravitate over time toward the undesirable all-weeds equilibrium, and thus are said to be in its basin-of-attraction. The plant states to the right are in the basin of attraction associated with a more desirable interior equilibrium including some level of grasses (case **(b)**), or the all-grasses equilibrium (case **(c)**). The stable manifold represents the threshold of environmental change referred to in the state-and-transition theory. The two conditions guaranteeing the existence of a threshold are:

- $q_w^u = q_w^l/B > 1$ (i.e., the all-weeds equilibrium is stable), and

- $N_g(g)$ and $N_w(g)$ intersect at least once in the phase plane.

5.   Analytical Approximation of the Stable Manifold

In order to determine the environmental threshold for a given grassland ecosystem, the stable manifold of the interior saddle-point equilibrium must be approximated with some accuracy. First, we know that the lower portion of the stable manifold has endpoints at the origin and the interior saddle point (call it X), whose coordinates depend on the choice of parameters. These endpoints allow us to make a first approximation to the stable manifold.

Exercise

8. Draw the line from the origin to equilibrium X on the computer-generated phase diagram from **Exercise 7**. How does this line compare to the curve you plotted earlier as an approximation to the stable manifold?

There is more information useful in this approximation. If the stable manifold is described as a curve in the g-w plane, that is, $w = W(g)$, then we know the value of $W(g)$ and its derivative at the equilibrium point X. In particular, let $X = (\gamma, \omega)$ and M be the matrix in the linearization of the system of equations **(4a)** and **(4b)** about the point X:

$$\begin{bmatrix} (g - \gamma)' \\ (w - \omega)' \end{bmatrix} = M \begin{bmatrix} (g - \gamma) \\ (w - \omega) \end{bmatrix}. \tag{6}$$

The matrix M has real eigenvalues of opposite sign, since X is a saddle point. Let the eigenvector associated with the negative eigenvalue be denoted (u, v), and assume that $u \neq 0$. Because the stable manifold is parallel to this eigenvector at the point X, it must have slope v/u there. Consequently, the second approximation to the stable manifold has the form $W(g) = ag + bg^2$, where a and b are chosen to satisfy the following conditions:

- $W(\gamma) = w$ (i.e., X must rest on the stable manifold); and

- $W'(\gamma) = v/u$ (i.e., the stable manifold must be tangent to the eigenvector at X).

This second approximation is a quadratic polynomial. One could, if necessary, obtain coefficients for an approximating polynomial of higher degree for the stable manifold. However, such a procedure is complicated and of limited effectiveness.

Exercise

9. Write the two equations allowing you to solve for the coefficients a and b in the approximation to the stable manifold. Using the same parameters as in **Exercise 7**, plot this curve on the same phase portrait from **Exercises 7** and **8**, and again compare it with a numerically computed approximation to the stable manifold. Now redo the exercise after changing the value of r_g to 0.4 and the value of r_w to 0.27. How does the approximation to the stable manifold for these new values compare with the numerically computed approximation?

Exercise 9 demonstrates that there is a problem with our approximation at the origin. The behavior of the stable manifold varies there according to the values for r_g and r_w. Whenever $r_w/r_g \neq 1$, there is a dominant direction for trajectories to leave a neighborhood of the origin. Specifically, when $r_w/r_g > 1$, trajectories tend to leave the origin tangent to the g-axis. Alternatively, when $r_w/r_g < 1$, trajectories tend to leave the origin tangent to the w-axis. It turns out that the stable manifold may be expanded in a series of the form:

$$W(g) = g^p \left[a_0 + a_1(g - \gamma) + a_2(g - \gamma)^2 + \cdots \right], \qquad (7)$$

where $p = r_w/r_g$ determines the behavior of the approximation as $g \to 0$. Our final approximation truncates this series after the first two terms:

$$W(g) = g^p \left[a_0 + a_1(g - \gamma) \right]. \qquad (8)$$

We continue to use the same conditions specified above to solve for the coefficients a_0 and a_1, that is, $W(\gamma) = w$ and $W'(\gamma) = v/u$.

Exercise

10. Write the two equations allowing you to solve for the coefficients a_0 and a_1 in the approximation to the stable manifold. Using the parameter values underlying **Exercise 7**, plot the approximation for the stable manifold given by **(8)** on the phase portrait from **Exercise 9**. How well does it agree with the curves that you computed earlier to approximate the separatrix? How do you expect the error in the approximation to behave?

6.   Discussion

We have applied Boyd's [1991] competition model of grassland ecosystems to develop a method for analytically approximating *successional thresholds*. Successional thresholds are stable manifolds in phase space that partition grassland conditions into basins of attraction gravitating toward socially desirable or socially undesirable plant states over time. A necessary condition for the existence of a successional threshold is that undesirable plant species are relatively strong competitors in colonizing grassland.

Successional thresholds provide a valuable management tool for monitoring the long-term resiliency of grassland in response to various human activities, principally livestock production. For example, overgrazing—of the native perennial grasses that livestock prefer—is generally identified as the culprit behind the successful invasion by less desirable annual grass species of millions of acres of grassland in the intermountain region of the United States. Much of this grassland is publicly owned by the United States and available to private citizens with public grazing leases. Federal land managers have had the responsibility of setting limits on the number of livestock grazing public land to ensure the land's *sustained yield* over time [Federal Land Policy and Management Act 43 U.S.C. §1732(a) (1982)]. Given that past grazing limits set by public managers have not arrested the invasion of less desirable grass species, managers are under increasing public pressure to impose further grazing reductions to reestablish the more desirable grass species.

Further reductions in grazing decrease the consumption of the perennial grasses, thereby increasing their competitive vigor vis-à-vis invading annual varieties. Our model indirectly accounts for this by reducing the value of the parameter E, since it is inversely related to the competitiveness of perennial grasses. Reducing the value of E tends to shift the successional threshold upward and to the left in the phase plane and consequently increases the size of the basin-of-attraction, leading to a successional plant state with some positive proportion of desirable perennial grasses. If the current plant state faced by the public manager is included in the increased portion of this basin of attraction, then further grazing restrictions should prove successful in redirecting grassland to a more desirable plant state (all other things being equal).

Exercise

11. Use the same parameters from **Exercise 7** and the approximated threshold from **Exercise 10**. Assume that the public grazing manager oversees grassland that is colonized 10% by perennial grasses and 70% by invading annual varieties, that is, $(g, w) = (0.1, 0.7)$. To which equilibrium will this plant state gravitate toward over time? Should the manager reduce the number of grazing livestock? What is the impact of a grazing reduction decreasing the value of E from 0.172 to 0.1?

7. Solutions to the Exercises

1. The equilibrium values w^e are 0 and 1. **Figure 1** gives a graph of w' vs. w. The solution to **2b)** is

$$w(t) = \frac{1}{1 + c_w e^{-rt}},$$

where c_w is a constant depending on initial conditions.

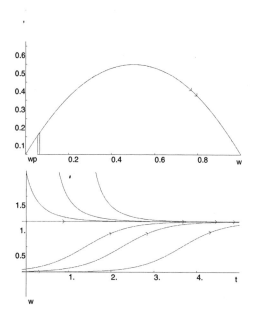

Figure 1. Plots of w' vs. w and of w vs. t for **Exercise 1**.

2. See **Figure 2**. The graph for q_g vs. g is qualitatively the same.

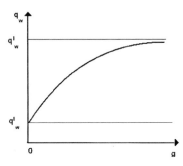

Figure 2. Solution for **Exercise 2**.

3.
$$N_g(g) = -\frac{g^2 + (BE - 1)g - BE}{q_w^l(E + g)},$$

$$N_w(g) = -\frac{q_g^u g^2 + \left(q_g^u BE - 1\right)g - E}{E + g}.$$

4. $E_c = \dfrac{q_g^u - 1}{1 - q_g^u B}.$

5. See **Figure 3**.

9.
$$a = \frac{1}{\gamma^2}\left(2\gamma\omega - \frac{v\gamma^2}{u}\right), \qquad b = \frac{1}{\gamma^2}\left(\frac{\gamma v}{u} - \omega\right).$$

For the parameters from **Exercise 7**, we have $a = 2.28033$, $b = 20.9499$.
For $r_g = 0.4$ and $r_w = 0.27$, we have $a = 6.05082$, $b = -2.77078$.

10.
$$a_0 = \frac{\omega}{\gamma^p}, \qquad a_1 = \frac{1}{\gamma^p}\left(\frac{v}{u} - \frac{p\omega}{\gamma}\right).$$

For the parameters from **Exercise 7**, we have $a_0 = 9.67505$, $a_1 = 8.13708$.

The error in approximation obtained by truncating the series after the a_1 term is given by $a_2 g^p(g - \gamma)^2$, where a_2 is in fact given as $W''(\alpha)/2$ at some point α between g and γ. While the second derivative of W is difficult to compute, the implications of this error term for the approximation are easily understood. First, it indicates that the approximation will be at its best near γ and near 0. One expects the approximation to be at its worst as g moves far to the right of γ. The approximation will be better as the power p increases, so that when the ratio r_w/r_g of intrinsic growth rates is small, the error in the approximation will be larger. It turns out that the second derivative of W is small, since most of the curvature is due to the term g^p. Thus, the error in the approximation is very small when $p > 1$ and is will within acceptable tolerances when $p < 1$.

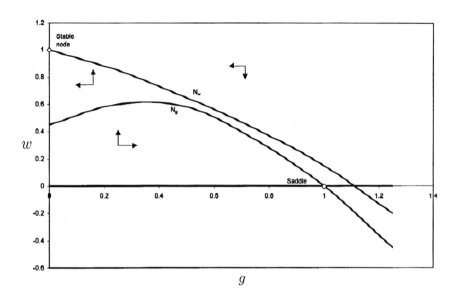

a. $E = 0.4$.

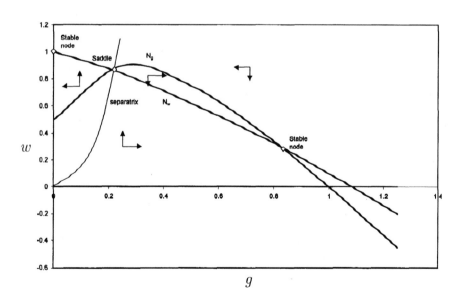

b. $E = 0.172$.

Figure 3ab. Solution for **Exercise 5ab**.

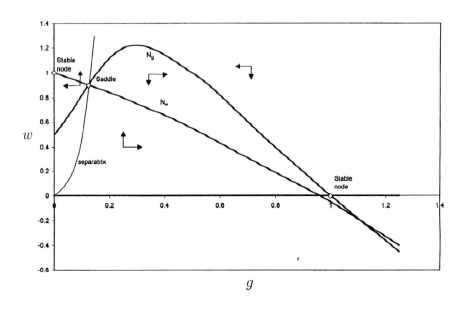

c. $E = 0.06$.

Figure 3c. Solution for **Exercise 5c**.

11. The initial plant state $(g, w) = (0.1, 0.7)$ is in the basin of attraction to the all-weeds equilibrium. If the grazing manager reduces the number of livestock so that the value of E declines from 0.172 to 0.1, the approximated threshold shifts upward and to the left (where $a_0 = 23.62$, $a_1 = 82.7589$). The initial plant state is now in the basin of attraction to a more desirable equilibrium containing some portion of native perennial grasses.

References

Boyd, E. 1991. A model for successional change in a rangeland ecosystem. *Natural Resources Modeling* 5(2): 161–189.

Evans, R., and J. Young. 1972. Competition within the grass community. In *The Biology and Utilization of Grasses* edited by V. Younger and C. McKell, 230–246. New York: Academic Press.

Hastings, A. 1996. *Population Biology: Concepts and Models.* New York: Springer-Verlag.

Laycock, W. 1991. Stable states and thresholds of range condition on North American rangelands: A viewpoint. *Journal of Range Management* 44(5): 427–433.

Stewart, G., and A. Hull. 1949. Cheatgrass (*Bromus tectorum*)—An ecological intruder in southern Idaho. *Ecology* 30(1): 58–74.

Westoby, M., B. Walker, and I. Noy-Meir. 1989. Opportunistic management for rangelands not at equilibrium. *Journal of Range Management* 42(4): 266–274.

About the Authors

Ray Huffaker is a natural resource economist specializing in mathematical bioeconomics.

Kevin Cooper is a mathematician specializing in numerical analysis and differential equations.

Thomas LoFaro is a mathematician whose area of expertise is dynamical systems.

UMAP

Modules in
Undergraduate
Mathematics
and Its
Applications

Published in
cooperation with

The Society for
Industrial and
Applied Mathematics,

The Mathematical
Association of America,

The National Council
of Teachers of
Mathematics,

The American
Mathematical
Association of
Two-Year Colleges,

The Institute for
Operations Research
and the Management
Sciences, and

The American
Statistical Association.

Module 776

Small Mammal Dispersion

Ray Huffaker
Kevin Cooper
Thomas Lofaro

**Applications of Calculus
to Biology and Ecology**

INTERMODULAR DESCRIPTION SHEET:	UMAP Unit 776
TITLE:	Small Mammal Dispersion
AUTHORS:	Ray Huffaker Department of Agricultural Economics huffaker@wsu.edu Kevin Cooper Thomas Lofaro Department of Pure and Applied Mathematics Washington State University Pullman, WA 99164
MATHEMATICAL FIELD:	Differential equations
APPLICATION FIELD:	Biology, ecology
TARGET AUDIENCE:	Students in a course in differential equations

ABSTRACT: This module introduces students to the social fence hypothesis explaining small mammal migration between adjacent land areas. Students are shown how the hypothesis is formulated in the population ecology literature as a pair of autonomous differential equations, and then they are directed toward a modified version of the standard formulation leading to increased realism. The modified version is solved qualitatively with phase diagrams for a range of ecological circumstances. Students also gain experience working with the numerical phase-plane plotter Dynasys, which can be downloaded from the World Wide Web. The social fence hypothesis is presented within the real-world context of controlling beaver-related damage in a given area by trapping.

PREREQUISITES: Introduction to ordinary differential equations covering phase-plane solutions.

Tools for Teaching 1999, 19–37. Reprinted from *The UMAP Journal* 20 (1) (1999) 47–65.
©Copyright 1999, 2000 by COMAP, Inc. All rights reserved.

COMAP, Inc., Suite 210, 57 Bedford Street, Lexington, MA 02420
(800) 77-COMAP = (800) 772-6627, or (781) 862-7878; http://www.comap.com

Small Mammal Dispersion

Ray Huffaker
Department of Agricultural Economics
huffaker@wsu.edu

Kevin Cooper
Thomas Lofaro
Department of Pure and Applied Mathematics
Washington State University
Pullman, WA 99164

Table of Contents

MODULES AND MONOGRAPHS IN UNDERGRADUATE
MATHEMATICS AND ITS APPLICATIONS (UMAP) PROJECT

The goal of UMAP is to develop, through a community of users and developers, a system of instructional modules in undergraduate mathematics and its applications, to be used to supplement existing courses and from which complete courses may eventually be built.

The Project was guided by a National Advisory Board of mathematicians, scientists, and educators. UMAP was funded by a grant from the National Science Foundation and now is supported by the Consortium for Mathematics and Its Applications (COMAP), Inc., a nonprofit corporation engaged in research and development in mathematics education.

Paul J. Campbell   Editor
Solomon Garfunkel   Executive Director, COMAP

1.   Introduction

Beavers (*Castor canadensis*) are the largest rodents in North America.[1] Adults are between three to four feet in length and weigh 30 to 60 pounds. Beavers possess a specialized digestive system that permits them to digest tree bark. They prefer the tender bark at the top of hardwood trees (i.e., maple, linden, birch, and poplar) and gain access to it by systematically gnawing at a tree's trunk until it falls over. Beavers also use felled trees to dam up slow moving streams and small rivers, thereby creating ponds that are an essential part of their habitat.

Beaver ponds create a wide range of potential environmental and agricultural benefits, including [Stuebner 1992]:

- habitat for a large number of wildlife species,

- improved water quality as sediment is allowed to settle out of turbid waters,

- improved productivity of adjacent land for livestock grazing, and

- fertile soils for agricultural production when drained.

Unfortunately, these potential benefits may come at what economists term an "opportunity cost." The opportunity cost of beaver ponds includes any environmental, aesthetic, or future commercial benefits trees were generating before being felled by beavers. Beavers may be deemed a public nuisance when the opportunity cost of their activities is thought to outweigh the benefits. In such case, the public might attempt to minimize beaver damage by controlling the population. For example, the North Carolina Legislature authorizes landowners to use any lawful method at any time to remove beavers destroying their property.

Trapping is generally the most effective means of controlling a beaver population whose primary damage in a given land area is felling trees. There have been many myopic attempts on a limited scale to trap and remove all beavers from a damaged area [Hill 1982]. However, experience from these attempted eradication efforts demonstrates that beavers from neighboring uncontrolled or lesser controlled areas tend to immigrate continually into the controlled area to fill the resulting population vacuum [Houston 1987].

Population ecologists have formulated the *social fence hypothesis* to explain how migration of small mammals between adjacent land areas might occur in general [Hestbeck 1982, 1988]. According to the hypothesis, individuals within a given area compete for vital resources, and when competition reaches a critical level, there is social pressure exerted on some individuals to depart (*within-group aggression*). For example, young beavers generally remain with the colony for only two years before departing to establish their own colonies.

---

[1]The information in this introduction was gleaned from the Web sites http://ngp.ngpc. state.ne.us/wildlife/beaver.html, http://www.educ.wsu.edu/enviroed/beavers.html, and http://www.ces.ncsu.edu/nreos/wild/beavers.html .

However, as individuals attempt to depart, there is territorial pressure exerted against their departure by neighboring populations (*between-group aggression*). When within-group aggression exerted in area A is stronger than between-group aggression exerted in neighboring area B, the social fence is said to be open for individuals to migrate from A to B.

A single-shot eradication effort has the unintended consequence of opening the social fence to migration from uncontrolled to controlled areas by removing beavers in the controlled area who otherwise would exert between-group aggression against potential immigrants. Consequently, sustained trapping efforts are required to offset this continual migration so that the controlled population can be maintained at a desired fixed level through time. The exercises below investigate the long-term impacts of a range of multi-period sustained trapping strategies on the population densities of beavers in neighboring controlled and uncontrolled areas. The exercises are based on the following mathematical formulation of the social fence hypothesis.

2. Mathematical Formulation of the Social Fence Hypothesis

Consider first the mathematical formulation of the social fence hypothesis in the absence of trapping. Let X and Y represent nonnegative beaver population densities (in beavers/square mile) in neighboring areas, and let $\dot{X}$ and $\dot{Y}$ represent the associated annual net rates of change (in beavers/square mile/year) according to:

$$\dot{X} = F_0(X)X - F_1(X, Y) \tag{1}$$

$$\dot{Y} = F_2(Y) + F_1(X, Y). \tag{2}$$

The net rate of change in annual population in each area is equal to the difference between the rates of net growth (i.e., birth rate minus the death rate) and dispersion. Functions $F_0(X)$ and $F_2(Y)$ are net proportional annual growth rates for X and Y, respectively, with units 1/year and are given by:

$$F_0(X) = R_X \left(1 - \frac{X}{K_X}\right) \tag{3}$$

$$F_2(X) = R_Y \left(1 - \frac{Y}{K_Y}\right), \tag{4}$$

where the quantities R_X (1/year), K_X (beavers/square mile), R_Y (1/year), and K_Y (beavers/square mile) are nonnegative constants. As the population density X approaches zero in its area, the net proportional growth rate approaches R_X (i.e., $F_0 \to R_X$ as $X \to 0$), which is called the *intrinsic growth rate* (1/year). Alternatively, as X approaches K_X, the net proportional growth rate decreases toward zero (i.e., $F_0 \to 0$ as $X \to K_X$) due to the negative impacts of crowding.

Thus, K_X is the environmental carrying capacity for beavers in X's area. The parameters R_Y and K_Y are interpreted analogously in Y's area.

Exercise

1. Assume that $F_1(X,Y)$ is zero in **(1)**, so that the population X changes only in response to net growth, that is, $\dot{X} = F_0(X)X$. Graph this differential equation and indicate with directional arrows how X changes over time, for the two cases when $0 < X < K_X$ and when $X > K_X$ (e.g., an arrow pointing rightward indicates that X increases over time). Next, solve the differential equation and graph the solution. Explain how the solution satisfies the limits on $F_0(X)$ as X approaches zero and the carrying capacity, respectively.

The total dispersion flux term, $F_1(X,Y)$ (beavers/square mile/year), is a mathematical representation of the social fence hypothesis found in the mathematical ecology literature [Stenseth 1988]. This literature assumes that whenever $X > Y$, the within-group aggression exerted by X is greater than between-group aggression exerted by Y, and the net proportional annual migration rate, $F_1(X,Y)$, acts as a dispersive valve allowing beavers to migrate from X to Y, that is, $F_1 > 0$. Alternatively, whenever $Y > X$, the dispersive valve opens in the opposite direction ($F_1 < 0$) for beavers to migrate from Y to X. In short, operation of the social fence is tied to the population differential in the two areas. A functional form for $F_1(X,Y)$ satisfying these assumptions is

$$F_1(X,Y) = B(X - Y)X, \tag{5}$$

where $B > 0$ is a constant parameter with units $(\text{year})^{-1}(\text{beaver/square mile})^{-1}$.

Exercise

2. Assume that the population density of Y is much greater than that of X, that is, let $Y = 100$ and $X = 1$. Calculate the total dispersion flux $B(X - Y)X$ in terms of B. Now assume that Y is only marginally larger than X, that is, let $Y = 100$ and $X = 99$, and recalculate $F_1(X,Y)X$. Compare the total dispersion fluxes for both cases and discuss. Does the operation of the social fence as formulated by **(5)** make ecological sense?

Another problem with the formulation of the social fence hypothesis in **(5)** is that it unreasonably triggers migration in the situation where $X > Y$ but X exerts much less pressure on its carrying capacity than Y because $K_X \gg K_Y$. Competition for vital resources is less keen for X than for Y, implying that within-group aggression pressuring emigration from X's area should be less than between-group aggression exerted against immigration into Y's area. Under these circumstances, the dispersive value should be closed to migration from X to Y. An alternative formulation of the social fence hypothesis rectifying the above problems is:

$$F_1(X, Y) = M \left(\frac{X}{K_X} - \frac{Y}{K_Y} \right), \tag{6}$$

where M is a constant rate with units (beavers/square mile/year). Whenever X constitutes a larger fraction of its carrying capacity than Y (i.e., $X/K_X > Y/K_Y$), within-group aggression of X is assumed to be greater than between-group aggression, and $F_1(X, Y)$ acts as a dispersive valve allowing beavers to migrate from X to Y, that is, $F_1 > 0$.

3.   The Social Fence Formulation with Trapping

Now assume that trapping occurs in X's area (i.e., the controlled area) but not in neighboring Y's area (i.e., the uncontrolled area). The modified rate equations are:

$$\dot{X} = F_0(X)X - F_1(X, Y) - PX \tag{7}$$
$$\dot{Y} = F_2(Y) + F_1(X, Y), \tag{8}$$

where $F_1(X, Y)$ is given by (6), P (1/year) represents the per capita annual trapping rate of X, and PX represents the total beavers trapped each year. As trapping reduces the population pressure on carrying capacity in X's area, between-group aggression exerted by X against migration from Y decreases (all other things being equal), and the dispersive valve may open for individuals to migrate from uncontrolled population Y to controlled population X.

4.   Dimensionless Rate Equations

The system of differential equations given by (7)–(8) can be simplified by making all variables and parameters dimensionless. Define dimensionless variables

$x = X/K_X$ (fraction of carrying capacity in the controlled area),

$y = Y/K_Y$ (fraction of carrying capacity in the uncontrolled area), and

$\tau = R_X t$ (dimensionless time variable).

Dimensionless parameters are

$m = M/R_X K_X$ (dimensionless dispersion parameter),

$p = P/R_X$ (dimensionless trapping parameter),

$r = R_Y/R_X$ (comparison of intrinsic growth rates in both areas), and

$k = K_X/K_Y$ (comparison of carrying capacities in both areas).

Exercise

3. Substitute these dimensionless quantities into **(7)–(8)** to show that the dimensionless model is:

$$x' = \frac{dx}{dt} = x(1 - x) - m(x - y) - px \tag{9}$$

$$y' = \frac{dy}{dt} = ry(1 - y) + km(x - y). \tag{10}$$

Note that the number of system parameters is reduced from six (R_X, R_Y, K_X, K_Y, M, P) to four (m, p, r, k).

We now analyze the solutions to the system **(9)–(10)** as the per capita trapping rate is increased from zero.

5.   Zero-Trapping Dynamics

Consider first the population dynamics of X and Y when no trapping occurs in the tree-damaged area where X resides (i.e., $p = 0$ in **(9)**).

Setting **(9)** equal to zero yields the following implicit expression for the $x' = 0$ nullcline, which we will call $N_x(y)$:

$$x^2 - (1 - m)x - my = 0. \tag{11}$$

The lack of an interaction term between x and y implies that $N_x(y)$ is a parabola in yx-space [Korn and Liberi 1978, 387] whose vertex occurs at a positive (negative) value of x when $(1 - m)$ is positive (negative). The downward (upward) sloping branch of the parabola intersects the origin when $(1 - m)$ is positive (negative).

Exercises

4. Set **(10)** equal to zero and solve for x in terms of y to derive the $y' = 0$ nullcline balancing net growth with diffusion each year in Y. Denote this nullcline by $N_y(y)$ and describe its graph.

5. Use the following baseline parameter values from a recent study [Huffaker et al. 1992] to plot the nullclines $N_x(y)$ and $N_y(y)$: $R_X = 0.335$, $R_Y = 0.3015$ ($r = 0.9$), $K_X = 1.107$, $K_Y = 0.9963$ ($k = 1.11$), and $M = 0.3473$ ($m = 0.937$).

6. The nullclines plotted in **Exercise 5** generate a dual steady-state configuration. One steady state occurs at the carrying capacities of X and Y (i.e., at $(x, y) = (1, 1)$) and the other occurs at the origin. Nullclines plotted with other parameter values can be shown to maintain these two steady states. The nullclines divide the phase plane into four regions called isosectors. Supply arrows indicating the directions of motion of x and y over time in each isosector.

7. Calculate the relevant eigenvalues to show that the equilibrium at the origin is a saddle point. Calculate the relevant eigenvalues to show that the equilibrium at the carrying capacities is a stable node. Draw in solution trajectories consistent with this information.

8. Use a computer program to generate numerically the phase diagram for the baseline parameters. One such program that you can download from the World Wide Web is DynaSys at http://www.sci.wsu.edu/idea/software. html .

Figure 1 is helpful in understanding the dynamics associated with various isosectors of the phase plane. The figure denotes the net proportional growth rate for x from **(9)** as $f_0 = (1 - x)$ and the net proportional growth rate for y from **(10)** as $f_2 = r(1 - y)$. Three dashed lines are superimposed on the nullclines from **Exercise 6** to divide phase space further into six regions. The regions bounded by the lines $x = 1$ and $y = 1$ (II and III) are characterized by positive net proportional growth rates for both populations, since each is below carrying capacity. Regions above $x = 1$ (I, IV, and V) produce negative net proportional growth rates for x, since the population is above carrying capacity. Regions to the right of $y = 1$ (IV, V, and VI) produce negative growth rates for y, since the population is above carrying capacity. The dashed line running from the origin through the nullclines at carrying capacity is the zero-dispersion line (zdl), which sets the dispersion flux term $f_1 = m(x - y)$ equal to zero since $x = y$. Population levels above the zdl (Regions I, II, and VI) open the social fence for migration from x to y, while levels below (Regions III, IV, and V) reverse the migratory flux.

Consider, for example, the dynamics in region II, which is bounded above by $x = 1$ and below by the zdl. Growth rates are positive for both populations, since each is below carrying capacity. The social fence is open for migration from x to y, since population levels are above the zdl. Thus, x enjoys a positive net proportional growth rate but suffers emigration losses. Initial levels of x above $N_x(y)$ initially decrease over time, because emigration losses are greater than growth each period. However, once x falls below $N_x(y)$, the growth rate overwhelms the emigration rate and the population begins to increase. Conversely, positive net proportional growth rates work together with immigration gains to increase y.

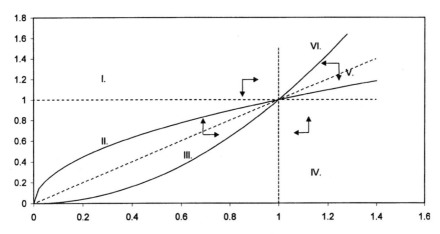

Figure 1. The zero-trapping plane ($p = 0$) with directions of motion from various isosectors. The plot is divided into regions by the dotted lines. Each region is labeled using roman numerals:

I.	$f_0 < 0, f_2 > 0$, beavers migrate from x to y;
II.	$f_0 > 0, f_2 > 0$, beavers migrate from x to y;
III.	$f_0 > 0, f_2 > 0$, beavers migrate from y to x;
IV.	$f_0 > 0, f_2 < 0$, beavers migrate from y to x;
V.	$f_0 < 0, f_2 < 0$, beavers migrate from y to x;
VI.	$f_0 < 0, f_2 < 0$, beavers migrate from x to y.

6.   Positive Trapping Dynamics

Consider now the impact of trapping some portion of x each year. The system of differential equations governing the evolution of neighboring beaver populations under these circumstances is **(9)–(10)**, with p set at fixed rate p_f. Assume that p_f represents a 100% annual trapping rate (i.e., $P = 1$ and $p = P/R_X = 2.985$) and that all other parameters are held at the baseline values given in **Exercise 5**. The nullcline for the uncontrolled population y remains the same as in the zero-trapping case.

Setting **(9)** equal to zero with $p = p_f$ yields the following implicit expression for the $x' = 0$ nullcline $N_x(y)$:

$$x^2 - (1 - m - p_f)\, x - my = 0. \tag{12}$$

The nullcline $N_x(y)$ is a parabola whose vertex occurs at a positive (negative) value of x when $(1 - m - p_f)$ is positive (negative). The downward (upward) sloping branch of the parabola intersects the origin when $(1-m-p_f)$ is positive (negative). Increasing the fixed trapping rate from zero shifts $N_x(y)$ downward. The nullcline $N_y(y)$ remains the same as in the zero-trapping case.

Exercises

9. Plot $N_x(y)$ and $N_y(y)$ for the baseline parameters and describe the resulting steady-state configuration. How does this configuration differ from that in the zero-trapping case? Does trapping open the social fence for sustained migration from y to x given the baseline parameter values?

10. Do the necessary eigenvalue analysis to determine the stability of all non-negative steady states, and draw in solution trajectories.

11. Use a computer program to generate the phase diagram for the baseline trapping scenario.

Table 1 shows the impact, on the positive steady-state levels of the controlled and uncontrolled populations, of the per capita trapping rate, P. In the table,

- X and Y are the dimensional controlled and uncontrolled population variables,

- x and y are their respective dimensionless counterparts measuring population pressure exerted on carrying capacity,

- $F_0(X)X$ and $F_2(Y)Y$ are their respective total sustained annual growth rates,

- $F_1(X,Y) < 0$ is sustained annual migration from Y to X, and

- PX is the total sustained annual trapping rate of X.

The annual growth functions are defined in **(3)–(4)** and the annual migration function is defined in **(5)**.

Table 1.

Effect of the per capita trapping rate P on population levels.

		P			
	0	0.25	0.5	0.75	1.0
X	0.072	0.610	0.295	0.141	0.072
Y	0.205	0.720	0.480	0.313	0.205
x					
y					
$F_0(X)X$					
$F_2(Y)Y$					
$F_1(X,Y)$					
PX					

Exercise

12. Do the required calculations to complete the table and generate the following plots using this information: First, plot x and y against P. Next, plot the total sustained annual trapping and migration rates against P. Finally, plot the total sustained annual growth rates of X and Y against P. Use these plots to explain why the total sustained annual trapping rate, PX, first increases and then decreases in response to increasing sustained per capita trapping rates, P.

7. Discussion

Landowners suffering beaver-related tree damage have several options. One extreme is to take no control action, which allows beavers to approach their environmental carrying capacity in the tree-damaged area. A landowner might be willing to forego control action if tree damage is outweighed by the many benefits provided by beaver ponds.

The other extreme is to attempt to eradicate beavers in the tree-damaged area with a single-shot trapping effort. This option may be futile because of the migratory behavior of neighboring beaver populations. The recently formulated social fence hypothesis explains the migratory behavior of small mammal populations as the ecological analogue of osmosis: Animals from a superior habitat are posited to diffuse through a social fence to a less densely populated habitat until the pressure to depart ("within-group aggression") is equalized with the pressure exerted against invasion ("between-group aggression"). According to this hypothesis, a landowner succeeding in removing the entire beaver population from an area in the short term unintentionally creates a population vacuum that is filled by migrating beavers from surrounding areas in the longer term. The specter of continual immigration into the controlled area justifies a sustained trapping strategy that offsets this sustained migration, so that the controlled population can be maintained at some fixed level through time.

We investigated the impact of sustaining various fixed per capita annual trapping rates. We applied those to steady-state beaver populations, in neighboring controlled and uncontrolled areas, by solving a mathematical model of the social fence hypothesis, using baseline parameters taken from a recent study. We found that:

- In the absence of trapping, both populations tend toward their respective carrying capacities.

- As the sustained per capita trapping rate increases from zero to relatively low levels, the steady-state population in the controlled area is driven to smaller fractions of its carrying capacity. This opens the social fence to increasing sustained migration from the uncontrolled to the controlled area and also—due to less crowding—increases the sustained net proportional growth rate in the controlled area. These additions to the population must be offset by trapping an increasing total number of beavers so that controlled population is sustainable at lower steady states. The steady-state population in the uncontrolled area is also driven to smaller fractions of its carrying capacity, due to emigration losses.

• As the sustained per capita trapping rate increases to relatively high levels, the steady-state populations in both areas continue to exert decreasing pressure on their respective carrying capacities. The total number of beavers that need to be trapped annually to sustain these decreasing steady-state levels declines, due to decreased migratory and growth pressures in the controlled area. Migratory pressure drops, because the social fence begins to close a bit, due to decreasing gaps between the pressures exerted on carrying capacity in the controlled and uncontrolled areas. Growth pressures drop, because the steady-state populations in the two areas decrease to the extent that their net proportional growth rates begin to decline.

In light of these results:

Should a landowner adopt a relatively low or high sustained per capita trapping rate?

The answer depends on the underlying biological and economic circumstances. Economists would expect a rational landowner to adopt a trapping rate that results a steady-state controlled population defined by the following characteristic:

Marginally decreasing the steady-state population by annually trapping one more beaver would generate sustained trapping costs outweighing sustained benefits measured as avoided tree damage.

All other things being equal, the economically optimal steady-state controlled population is expected to be relatively low (reflecting a relatively high sustained per capita trapping rate) when beavers cause significantly more tree damage than they cost to trap, even at low population levels. See Huffaker et al. [1992] for an extended discussion of this question.

8.   Solutions to the Exercises

1. See **Figure 2**. The solution of the differential equation is

$$X(t) = \frac{1}{1 + ce^{-rt}},$$

where c is a constant depending on the initial conditions; this function is graphed in **Figure 3**.

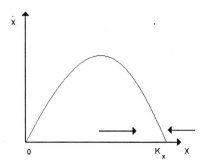

Figure 2. Graph of **(1)** when $F_1(X) = 0$.

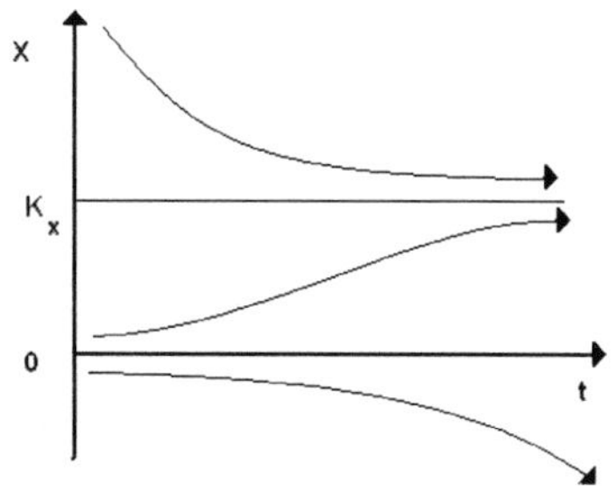

Figure 3. Graph of solutions to **(1)** for various initial conditions.

2. For $Y = 100$, $X = 1$: $F_1(X,Y) = B(1 - 100)(1) = -99B$.
For $Y = 100$, $X = 99$: $F_1(X,Y) = B(99 - 100)(99) = -99B$.
  The problem with the formulation of the social fence in **(5)** is that it can generate the above result, where the same number of beavers migrate annually from Y to X regardless of whether the population differential between the two is large or small.

3. $N_y(y) = \dfrac{1}{km}\left[ry^2 - (r - km)y\right]$.

See **Figure 4** for graphs for the two cases $r - km > 0$ and $r - km < 0$.

6. See **Figure 5**.

7. See **Figure 6**. The eigenvalues are:
for $(x, y) = (0, 0)$: $0.953864, -1.03093$;
for $(x, y) = (1, 1)$: $-0.951343, -2.92573$.

9. The origin remains a steady-state solution, but trapping drives the interior steady state below carrying capacity for both populations. Trapping opens the door to sustained migration from Y to X, given the baseline parameters.

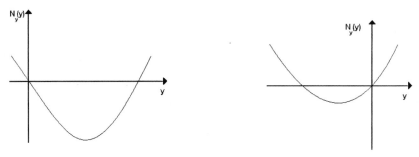

Figure 4. Graph of $N_y(y)$ for **a)** $r - km > 0$. **b)** $r - km < 0$.

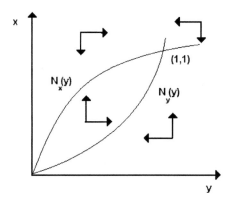

Figure 5. Solution to **Exercise 6**.

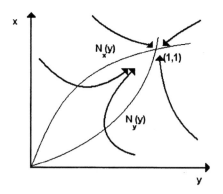

Figure 6. Solution to **Exercise 7**.

10. See **Figure 7**. The eigenvalues are:
    for $(x, y) = (0, 0)$: $0.174641, -3.22671$;
    for $(x, y) = (0.21, 0.061)$: $-0.177817, -3.38225$.

12. See **Table 2** and **Figure 8**. See the **Discussion** section of the Module for explanation of why the total sustained annual trapping rate, PX, first increases and then decreases in response to increasing sustained per capita trapping rate, P.

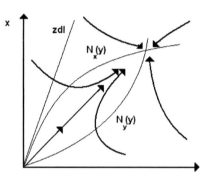

Figure 7. Solution to **Exercise 10**.

Table 2.

Effect of P on population levels.

P	0	0.25	0.5	0.75	1.0
X	0.072	0.610	0.295	0.141	0.072
Y	0.205	0.720	0.480	0.313	0.205
x	1	0.550	0.270	0.128	0.065
y	1	0.720	0.480	0.310	0.210
$F_0(X)X$	0	0.092	0.072	0.041	0.022
$F_2(Y)Y$	0	0.060	0.075	0.065	0.049
$F_1(X,Y)$	0	-0.060	-0.080	-0.060	-0.050
PX	0	0.152	0.147	0.106	0.072

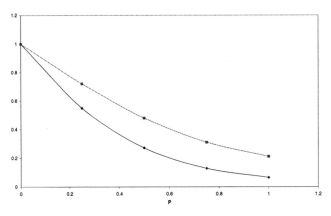

a) Graphs of y (upper) and x (lower) vs. P.

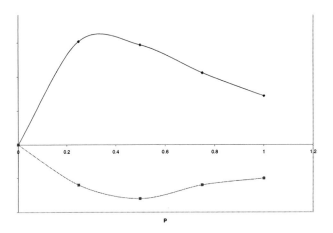

b) Graphs of PX (upper) and $F_1(X, Y)$ (lower) vs. P.

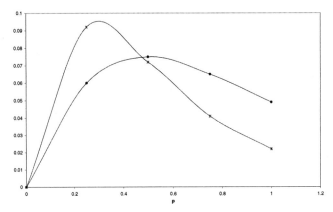

c) Graphs of $F_0(X)X$ (lower at $P = 1$) and $F_2(Y)Y$ (higher at $P = 1$) vs. P.

Figure 8. Solution to **Exercise 12**.

References

Hestbeck, J. 1982. Populations regulation of cyclic mammals: The social fence hypothesis. *Oikos* 39 (1982):157–163.

_____. 1988. Population regulation of cyclic mammals: A model of the social fence hypothesis. *Oikos* 52 (1988) (2):156–168.

Hill, E.P. 1982. The beaver. In *Wild Mammals of North America: Biology, Management, and Economics,* edited by J.A. Chapman and G.A. Feldhamer. Baltimore, MD: Johns Hopkins University Press.

Houston, A. 1987. Beaver control and reclamation of beaver-kill sites with planted hardwoods. Unpublished. Knoxville, TN: University of Tennessee.

Huffaker, R., M.G. Bhat, and S.M. Lenhart. 1992. Optimal trapping strategies for diffusing nuisance-beaver populations. *Natural Resource Modeling* 6 (1992) (1): 71–97.

Korn, H., and A. Liberi. 1978. *An Elementary Approach to Functions.* 2nd ed. New York: McGraw-Hill.

Stenseth, N. 1988. The social fence hypothesis: A critique. *Oikos* 52 (1988) (2): 169–177.

Stuebner, S. 1992. Leave it to the beaver. *High Country News* 24 (1992) (15): 1.

About the Authors

Ray Huffaker is a natural resource economist specializing in mathematical bioeconomics.

Kevin Cooper is a mathematician specializing in numerical analysis and differential equations.

Thomas LoFaro is a mathematician whose area of expertise is dynamical systems.

UMAP

Modules in Undergraduate Mathematics and Its Applications

Published in cooperation with

The Society for Industrial and Applied Mathematics,

The Mathematical Association of America,

The National Council of Teachers of Mathematics,

The American Mathematical Association of Two-Year Colleges,

The Institute for Operations Research and the Management Sciences, and

The American Statistical Association.

Module 777

Microcosm Macrocosm: Population Models in Biology and Demography

George Ashline and
Joanna Ellis-Monaghan

**Applications of Calculus
to Demography and Biology**

COMAP, Inc., Suite 210, 57 Bedford Street, Lexington, MA 02420 (781) 862–7878

INTERMODULAR DESCRIPTION SHEET: UMAP Unit 777

TITLE: Population Models in Biology and Demography

AUTHOR: George Ashline
                                    Joanna Ellis-Monaghan
                                    Dept. of Mathematics
                                    St. Michael's College
                                    Winooski Park
                                    Colchester, VT 05439
                                    gashline@smcvt.edu
                                    jellis-monaghan@smcvt.edu

MATHEMATICAL FIELD: Calculus

APPLICATION FIELD: Demography, biology

TARGET AUDIENCE: Students in first-semester calculus.

ABSTRACT: This module presents applications from microbiology and demography that give a physical context for critical concepts in a first-semester calculus course. Elementary models of population growth are developed using data collected from biological experiments conducted during the course. Then, proceeding from "microcosm to macrocosm," the tools of calculus, demographic software, and the Maple computer algebra system (CAS) are used to address questions of human population projections. Included are discussion materials, sample models using exponential and logistic curves, projects, complete biology labs for generating raw data, exercises for developing facility with Maple, and resource lists (including software and Internet sites) for studying demographic questions.

PREREQUISITES: None.

Population Models in Biology and Demography

George Ashline
Joanna Ellis-Monaghan
Dept. of Mathematics
St. Michael's College
Winooski Park
Colchester, VT 05439
gashline@smcvt.edu
jellis-monaghan@smcvt.edu

Table of Contents

MODULES AND MONOGRAPHS IN UNDERGRADUATE MATHEMATICS AND ITS APPLICATIONS (UMAP) PROJECT

The goal of UMAP is to develop, through a community of users and developers, a system of instructional modules in undergraduate mathematics and its applications, to be used to supplement existing courses and from which complete courses may eventually be built.

The Project was guided by a National Advisory Board of mathematicians, scientists, and educators. UMAP was funded by a grant from the National Science Foundation and now is supported by the Consortium for Mathematics and Its Applications (COMAP), Inc., a nonprofit corporation engaged in research and development in mathematics education.

Paul J. Campbell   Editor
Solomon Garfunkel   Executive Director, COMAP

1.   Introduction

The main objective of this Module is to use biology and demography to provide interesting, motivating contexts for the material in a first-semester calculus course. The activities demonstrate how a mathematical model can address complex questions and yield answers simply not discernible from the raw data alone. This connection strongly motivates the use of modeling and the importance of mathematics in studying biological and sociological issues (and, by extension, issues in other fields).

The Module begins by analyzing biological growth and concludes with a major research project that investigates how laboratory results can be extended to consider the question of human population change. Proceeding from micro-biology to demography demonstrates how laboratory experiments can serve as a microcosm for larger questions. The activities give experience in the process of questioning and validating mathematical techniques. They also give exposure to some of the inherent difficulties of modeling population changes.

The extremes resulting from choosing an inappropriate model for human population change can be seen in the following rather lurid scene from the 1973 dystopian science fiction film *Soylent Green* [Fleisher 1973]: The year is 2022 A.D. Forty million people crowd New York City. An air-locked plastic bubble preserves the small handful of sickly trees remaining. Bodies pack stairwells at night. Only the very wealthy can afford fresh vegetables, strawberry jam, or even hot running water. Teeming masses of desperate people mill around derelict cars on filthy streets, wearing surgical masks against the thick yellow smog. They fight for crumbs of government-supplied protein crackers, and bulldozers control daily food riots, scooping crying, starving people into bloody piles. Sanitation trucks arrive for routine collection of the dead.

Is this be a realistic vision of the future or not? Who would predict such population growth and on what basis? In the prologue to the book *Make Room! Make Room!* that was the basis for *Soylent Green*, author Harry Harrison writes that

> [W]ithin fifteen years, at the present rate of growth, the United States will be consuming over 83 per cent of the annual output of the earth's materials. By the end of the century, should our population continue to increase at the same rate, this country will need more than 100 per cent of the planet's resources to maintain our current living standards. This is a mathematical impossibility—aside from the fact that there will be about seven billion people on this earth at that time and—perhaps—they would like to have some of the raw materials too.   [Harrison 1966, Prologue]

Now, more than 30 years later, Harrison's prediction has clearly been disproved. His statement shows that he based his vision of the future on an exponential model of human population growth. Throughout history, many people have chosen, either deliberately or out of ignorance of other models, to use a particular population model specifically to make a case for a particular point of view. For example, Paul Ehrlich used an exponential model to predict that in 900 years world's population will be 60,000,000,000,000,000, or 100 people per square yard of the earth's surface, land, and sea [Ehrlich 1968, 18]. He then used this dire prediction to justify such draconian measures as involuntary sterilization of the general population through contraceptives in the water supply, with the antidote "carefully rationed by the government to produce the desired population size" [Ehrlich 1968, 135], although he admitted such measures would be difficult to enforce.

In fact, not only are populations in many parts of the world not growing exponentially, they aren't even growing—they are declining. The replacement growth rate (the rate needed for an unchanging population) is about 2.2 children born to each woman [Cohen 1995, 288]. However, in 1997 the rate in both Spain and Italy was only 1.2 children per woman. Europe as a whole is well below replacement level, with an average total fertility rate of 1.4. Even the United States is at only 2.0 children per woman [Population Reference Bureau 1999]. Many other countries have fertility levels below the replacement level, and it is expected that this trend will also eventually manifest itself in many of the remaining countries [Tapinos and Piotrow 1980, 168–169].

So, although exponential models can be very helpful if used appropriately (and these models are developed in this Module), as Cohen points out, they have repeatedly failed as good forecasters of long-term population growth:

> Because of its great simplicity, the exponential model is remarkably useful for very short-term predictions: the growth rate of a large population during the next one to five years usually resembles the growth rate of that population over the past one to five years. Because of its great simplicity, the exponential model is not very useful for long term predictions, beyond a decade or two. Surprisingly, in spite of the abundant data to the contrary, many people believe that the human population grows exponentially. It probably never has and probably never will.   [Cohen 1995, 84]

Why is population change such an urgent concern? Who wants to know? And why is it so hard to predict? Carl Haub addresses the first two questions:

> Interest in [population] projections involves much more than a simple curiosity about what may lie ahead. Having some sense of the number of people expected, their age distribution, and where they will be living provides city planners and local governments, for instance, sufficient "lead time" to prepare for coming needs in terms of schools and traffic lights, or reservoirs and pipes to deliver water supplies. Businesses have a vital interest in the coming demand for their products and services.
> [Haub 1987, 3]

World population projections have been a cause of considerable concern, particularly because of the population "explosion" of the post–World War II years. The probable consequences of rapidly expanding human numbers have been the subject of lively debate. Recently, that debate has been joined in controversy by the issue of population decline resulting from the very low birth rates in some countries. Books on these concerns range from Ehrlich's *The Population Bomb* [1968] to Ben Wattenberg's *The Birth Dearth* [1987].

Despite the urgency of the questions, answers to human population concerns are hard to determine. Complicating the prediction process is the fact that the forecasters may themselves be biased, influencing their choice of models and their basic assumptions. As Julian Simon notes,

[S]everal groups have had a parochial self-interest in promoting these doom-saying ideas [of explosive population growth leading to scarce resources]. . . . [T]hese groups include (a) the media, for whom impending scarcities make dramatic news; (b) the scientific community, for whom fears about impending scarcities lead to support for research that ostensibly will ease such scarcities; and (c) those political groups that work toward more government intervention in the economy; supposedly worsening scarcities provide an argument in favor of such intervention.

[Simon 1990, 3–4]

Demographers honestly admit that they cannot predict the future: "[M]ost professional demographers no longer believe they can predict precisely the future growth rate, size, composition and spatial distribution of populations" [Cohen 1995, 109–110]. Furthermore, any mathematical model hoping to predict future conditions can only assume that existing trends will continue unchanged or will change in a predictable way. This is a very tenuous assumption, especially regarding such factors as fertility and mortality rates.

People themselves create the major difficulty in applying these models to the human condition. Unlike bacteria in a laboratory which simply reproduce until they have exhausted all available nutrients, humans have highly complex value systems which effect their reproductive trends and use of resources.

According to Cohen,

Humans seem to resolve conflicts of values by personal and social processes that are poorly understood and virtually unpredictable at present. How such conflicts are resolved can materially affect human carrying capacity, and so there is a large element of choice and uncertainty in human carrying capacity. . . . Not all of those choices are free choices. Natural constraints restrict the possible options . . . . [Cohen 1995, 296]

Because of this,

Estimating how many people the Earth can support requires more than demographic arithmetic. . . . [I]t involves both natural constraints that humans cannot change and do not fully understand, and human choices that are yet to be made by this and by future generations. Therefore the question "how many people can the Earth support?" has no single numerical answer, now or ever. Because the Earth's human carrying capacity is

3

constrained by facts of nature, human choices about the Earth's human carrying capacity are not entirely free, and many have consequences that are not entirely predictable. Because of the important roles of human choices, natural constraints and uncertainty, estimates of human carrying capacity cannot aspire to be more than conditional and probable estimates
. . . .   [Cohen 1995, 261–262]

But decisions must be made by the business community, by governments, by policy makers, and by individuals. These decisions depend on estimates of future conditions. So how can population trends be predicted, to allocate resources for example? And how is the choice of model justified and its effectiveness evaluated? To address these questions, it is first necessary to know what models are available and why they might be applied to human population changes. It is also necessary to acquire the mathematics to evaluate and manipulate these models. The first-semester calculus course develops the necessary mathematical tools to understand exponential and logistic curves. These curves can be used as models to analyze the growth patterns demonstrated in the laboratory component of this module. The collection of data from laboratory cultures of various organisms shows that both functions are quite effective models of population growth. In fact, both can even give quite accurate predictions of future growth.

So, it seems reasonable to apply these models to the human populations of various countries, and the culminating project of this Module provides the opportunity to do just that. Human population data can be collected from a variety of sources, including Internet sites giving the most up-to-date information currently available. Short-term predictions can be compared to known results, and the new data can be used to refine the models. Finally, the models can be used to predict future populations, both for individual countries and for the entire world population. Internet resources are available to see how these predictions compare to the most recent estimates of professional demographers. For most areas of the world, these models prove especially good for backward analysis and passable for very short-term predictions of future change. However, they usually have limitations as long-range forecasters of human populations. References to other models that require more advanced mathematical techniques are given in Sections 5.2–5.3 for those who are interested in pursuing these questions.

2.   Modeling Examples

We develop models for a bacteria colony in a controlled setting and for the population of Italy.

2.1 Raw Data

2.1.1 Bacteria Colony

We present data for a bacteria culture in the spring of 1995. The bacteria were grown in a petri dish, and every few days the diameter of the bacteria culture was measured. **Table 1** gives the recorded data points, where t_i is the time (in hours, converted to a decimal, after 12:45 P.M. on 4/11/95), d_i is the diameter (in millimeters), and $a_i = \pi d_i^2/4$ is the area covered by the colony. **Figure 1** shows the area as a function of time.

Table 1.
Data for a bacteria culture.

Index	Time (hours)	Diameter (mm)	Area (mm^2)
0	0	2.5	4.91
1	23.17	5	19.63
2	49.92	5	19.63
3	98.5	6	28.27
4	148.85	6.6	34.21
5	167.5	6.75	35.78
6	192.17	6.52	33.39

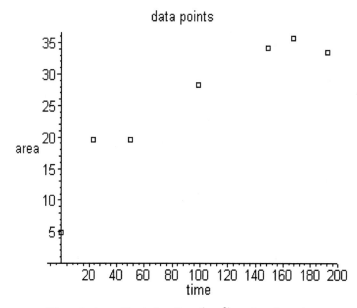

Figure 1. Area of bacteria culture (mm^2) vs. time (hours).

2.1.2   Population of Italy

The census data from Italy in this second example interestingly shows a similar growth pattern. The similarity suggests that a model developed for bacteria growth might be applicable to human population. **Table 2** and **Figure 2** give the population of Italy for several years prior to 1950, taken from various editions of *World Almanac and Book of Facts*.

Table 2.

The population of Italy.

Date	Population
1915	35,240,000
1921	37,270,493
1928	41,168,000
1931	42,118,835
1936	42,527,561
1940	45,330,441
1943	45,801,000
1946	45,646,000
1948	45,706,000

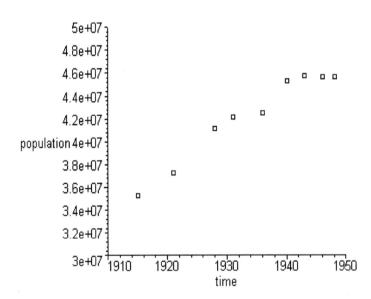

Figure 2. The population of Italy.

6

2.2 Exponential and Logistic Functions

To model data such as those in the examples above, it is necessary to be familiar with functions whose graphs are similar to the shapes displayed by the data. The following functions are useful for modeling certain kinds of growth, including the examples of bacteria growth and the population of Italy. The functions and graphics below, and those in the following section of modeling examples, are given using Maple, a computer algebra system. Section 4, **Developing Facility with Maple**, contains more information and examples of Maple code.

Exponential growth: The exponential function $y(t) = Ce^{kt}$, where C is the initial amount and k is the growth constant, models unrestricted growth with the rate of growth proportional to the amount present. Thus, the function satisfies the differential equation $y' = ky$. For example, **Figure 3** shows a plot of $y(t)$, with $C = 5.05$ and $k = .023$, generated from the following Maple code:

```
> y:=t->C*exp (k*t):
> C:=5.05:   k:=0.023:
> plot(y(t), t=-50..100);
```

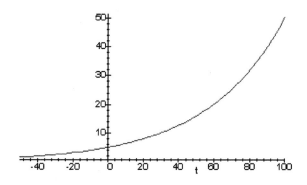

Figure 3. Exponential growth.

Limited growth: The limited growth curve models bounded growth, where the rate of growth is proportional to how close the amount is to the carrying capacity of the system. Such a curve looks like $y(t) = M(1 - e^{-kt})$, where M is the carrying capacity and k is the growth constant; it satisfies the differential equation $y' = k(M - y)$. **Figure 4** shows a plot of $y(t)$, with $M = 15$ and $k = 1.23$, generated from the following Maple code:

```
> y:=t->M*(1-exp(-k*t)):
> M:=15:   k:=1.23:
> plot(y(t), t=-1..4, y=-10..18);
```

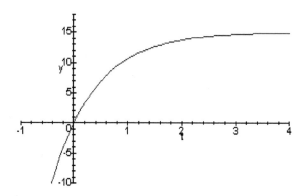

Figure 4. Limited growth.

Logistic growth: The logistic curve models growth where the rate of growth is proportional to both the amount present and the difference between the carrying capacity and the amount present. A logistic function looks like $y(t) = M/(1 + Be^{-Mkt})$, where M is the carrying capacity, B is a number that depends on the carrying capacity and the initial amount, and k is the growth constant; it satisfies the differential equation $y' = ky(M - y)$. **Figure 5** shows a plot of $y(t)$, with $M = 25$, $B = 22$, and $k = 0.23$, generated from the following Maple code:.

```
> y:=t->M/(1+B*exp(-M*k*t)):
> M:=25:  B:=22:  k:=0.23:
> plot(y(t), t=-1..3);
```

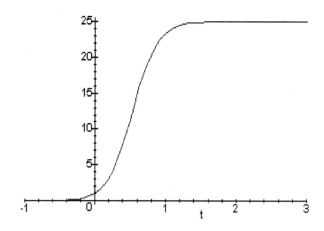

Figure 5. Logistic growth.

8

2.3 Modeling Bacteria Growth

We use a logistic curve to model the growth of the bacteria culture. We enter the raw data from **Table 1** into Maple:

```
> s:=evalf(12+45/60):  t[0]:=0:  d[0]:=2.5:
        t[1]:=23.0+10/60:  d[1]:=5:  t[2]:=48.0+2-5/60:  d[2]:=5:
        t[3]:=4*24+2.5:  d[3]:=6:  t[4]:=6*24+4.0+51/60:  d[4]:=6.6:
        t[5]:=7*24-.5:  d[5]:=6.75:  t[6]:=8*24.0+10/60:  d[6]:=6.52:
```

Then we compute the areas a_i using a loop

```
> for i from 0 to 6 do
>         a[i]:=Pi*.25*d[i]^ 2
> od:
```

Pairs of data points, together with an estimate for the carrying capacity M, are used to determine a logistic function $L(t)$ that models the growth of the bacteria. The general form of $L(t)$ is:

```
> L:=t-> M/(1+B*exp(-M*k*t));
```

$$L(t) = \frac{M}{1 + Be^{-Mkt}}$$

A good way to get a preliminary estimate for M is to plot the data points and do a freehand sketch of an "S" shape through them. Descriptions of more-refined techniques for determining the values of the parameters can be found in the articles on curve-fitting in Section 5.3.

We get the plot of **Figure 6** by inputting:

```
> with(plots):
> A:=plot([ [t[0],a[0]], [t[1],a[1]], [t[2],a[2]], [t[3],a[3]],
            [t[4],a[4]], [t[5],a[5]], [t[6],a[6]] ],
           x=-10..200, y=0..a[5]+1, style=point,
           title='data points', symbol=box, labels=[time,area]):
> display(A);
```

The second data point looks a little too high, and the last one a little too low, probably due to inaccuracy in measurement. A good guess for the carrying capacity M seems to be 34.5, so now $L(t)$ looks like:

```
> M:=34.5:  L(t);
```

$$L(t) = \frac{34.5}{1 + Be^{-34.5kt}}$$

Now, using t_0 and a_0 to solve for B gives:

9

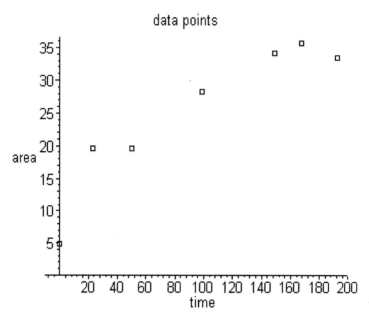

Figure 6. Logistic fit to bacteria growth.

```
> solve(a[0]=L(t[0]),B): B:=%;
```
$$B = 6.028282287$$

Thus, $L(t)$ becomes:

```
> L(t);
```
$$L(t) = \frac{34.5}{1 + 6.028282287e^{-34.5kt}}$$

Next, t_2 and a_2 are used to find k; the data point (t_1, a_1) looks as though it might be from a bad measurement, so (t_2, a_2) will probably yield a better model:

```
> solve(a[2]=L(t[2]),k);
```
$$0.001204767896$$

```
> k:=%;
```

With this, $L(t)$ becomes:

```
> L(t);
```
$$L(t) = \frac{34.5}{1 + 6.028282287e^{-0.0415649241t}}$$

Now the model can be compared to the full data set. Here are the values of the function $L(t)$ and the actual areas at the times t_0 through t_6:

```
> seq(evalf(L(t[i])), i=0..6);
```

4.908738521, 10.44986633, 19.63495408, 31.34950928, 34.07758738, 34.30413399, 34.42948532

```
> seq(evalf(a[i]), i=0..6);
```

4.908738522, 19.63495409, 19.63495409, 28.27433389, 34.21194400, 35.78470382, 33.38759009

All but the second are fairly close. The following command generates **Figure 7**, which shows how the actual data points fit on the curve:

```
> BC := plot(L(t), t=0..200, y=0..M+1, labels=[time,area],
             title='L(t) and data points'):
> display(A,BC);
```

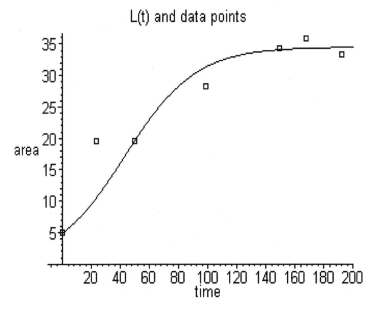

Figure 7. The data points and the model.

The most important aspect of having developed a model is that it is now possible to answer some questions that could not have been determined just using the observed data.

- *What was the area covered by the colony at 10:00 A.M. on 4/16?*

First, it is necessary to determine the value of t at that time by counting 5

11

days less 2:45 hours from the start of the experiment at 12:45 on 4/11. This gives:

```
> t[S]:=5*24-2.75;
```
$$t_s = 117.25$$

Now, evaluating $L(t)$ at that time gives the area covered:

```
> L(t[S]):
```
$$32.97968560$$

- *At what rate is the area increasing at that time?*

The derivative of $L(t)$ is needed to answer this. Fortunately, $L(t)$ satisfies the differential equation $y' = ky(M - y)$, which gives the following formula for $L'(t)$:

```
> Lprime:=t-> k*(L(t))*(M-L(t));
```
$$Lprime := t \rightarrow kL(t)(M - L(t))$$

Evaluate at t_s to find the rate:

```
> Lprime(t[S]);
```
$$.06040644899$$

- *What is the carrying capacity of the system?*

The carrying capacity is just M.

```
> M;
```
$$34.5$$

- *When does the rate slow down? In other words, when does the population stop growing at an increasing rate and start growing at a decreasing rate? (This just asks where the inflection point is.)*

A good estimate for this can be found by inspecting the graph: about 43 hours after the start of the experiment. Alternatively, the second derivative of $L(t)$ can be set equal to zero, and solving for t gives the t-coordinate of the inflection point:

```
> diff(L(t),t$2):
> solve(%=0,t);
```
$$43.22107658$$

- *What is the average area covered by the colony during the course of the experiment?*

First, integrating $L(t)$ from t_0 to t_6 gives the total area:

```
> int(L(t), t=t[0]..t[6]);
```
$$5012.92708$$

Then, dividing this by $t_6 - t_0 = t_6$ yields the average area (in mm^2) covered by the colony during the course of the experiment:

```
> int(L(t),t=t[0]..t[6])/t[6];
```
$$26.08635080$$

2.4   Modeling the Population of Italy

We use a logistic curve to model the population of Italy from 1915 to 1948, then consider how well the model predicts the population later.

We input the population of Italy for the nine years shown in the chart in Section 2.1, as well as for six later dates:

```
> restart;
> t[0]:=1915:   p[0]:=35240000:   t[1]:=1921:   p[1]:=37270493:
  t[2]:=1928:   p[2]:=41168000:   t[3]:=1931:   p[3]:=42118835:
  t[4]:=1936:   p[4]:=42527561:   t[5]:=1940:   p[5]:=45330441:
  t[6]:=1943:   p[6]:=45801000:   t[7]:=1946:   p[7]:=45646000:
  t[8]:=1948:   p[8]:=45706000:   t[9]:=1960:   p[9]:=50763000:
  t[10]:=1965:   p[10]:=52736000:   t[11]:=1970:   p[11]:=53670000:
  t[12]:=1975:   p[12]:=55810000:   t[13]:=1980:   p[13]:=57040000:
  t[14]:=1990:   p[14]:=57657000:
```

We use two data points and an estimate for M to create a logistic function $L(t)$ to model the population of Italy with the pre-1950 population data. A logistic curve with 1915 as its initial date has the following form, for some constants M, B, k:

```
> L:=t-> M/(1+B*exp(-M*k*(t-1915)));
```
$$L(t) = \frac{M}{1 + Be^{-Mk(t-1915)}}$$

As before, M can be estimated from a plot (**Figure 8**):

```
> with(plots):
> A:=plot([[t[0],p[0]], [t[1],p[1]], [t[2],p[2]], [t[3],p[3]],
          [t[4],p[4]], [t[5],p[5]], [t[6],p[6]], [t[7],p[7]],
          [t[8],p[8]]], x=1900..2000, y=0..60000000,
```

```
          style=point, title='data points',
          symbol=box, labels=[time, population]):
> display(A);
```

data points

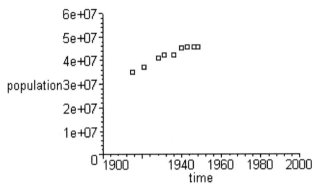

Figure 8. Estimating M from a plot.

A good guess for M from these points seems to be 50 million, so that $L(t)$ becomes:

```
> M:=50000000:  L(t);
```

$$L(t) = \frac{50000000}{1 + Be^{-50000000k(t-1915)}}$$

Next, using t_0 and p_0 to solve for B gives:

```
> solve(p[0]=L(t[0]),B):
> B:=evalf(%);
```

$$B = .4188422247$$

Now L(t) looks like:

```
> L(t);
```

$$L(t) = \frac{50000000}{1 + .4188422247e^{-50000000k(t-1915)}}$$

In turn, t_5 and p_5 are used to find k. This data point is used since it is part of a group of points toward the end of the modeling period:

```
> solve(p[5]=L(t[5]),k);
```

$$.1122122542 \ 10^{-8}$$

```
> k:=%:
```

14

Thus, after simplifying, $L(t)$ looks like:

```
> L(t);
```

$$L(t) = \frac{50000000}{1 + .4188422247e^{-.05610612710t+107.4432e34}}$$

The values of the function $L(t)$ at the times t_0 through t_8 and the actual populations prior to 1950 are:

```
> seq(evalf(L(t[i])), i=0..8);
```

.3523999999 10^8, .3848747902 10^8, .4159840763 10^8, .4271018355 10^8,
.4428974587 10^8, .4533044101 10^8, .4599589675 10^8, .4657368685 10^8,
.4691493452 10^8

```
> seq(evalf(p[i]), i=0..8);
```

.35240000 10^8, .37270493 10^8, .41168000 10^8, .42118835 10^8,
.42527561 10^8, .45330441 10^8, .45801000 10^8, .45646000 10^8,
.45706000 10^8

Thus, the modeling function seems fairly accurate up to 1950. **Figure 9** shows how the data points fit this curve:

```
> BC:=plot(L(t),t=1900..2000,y=0..60000000,
           labels=[time,population], color=blue,
           title='L(t) and data points'):
> display(A,BC);
```

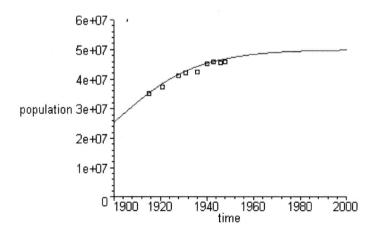

Figure 9. The data points and the model.

15

The model can now be used to predict the population of Italy in 1960, 1970, and 1980, which we can compare with population data published in the *World Almanac*:

```
> L(1960),L(1970),L(1980);
```
$$.4837745298 \ 10^8, .4906108639 \ 10^8, .4945989787 \ 10^8$$
```
> p[9],p[11],p[13];
```
$$50763000, \ 53670000, \ 57040000$$

Thus, this first model is not growing at a fast enough rate to reflect later population growth. However, the new data from 1970, 1980, and 1990 can be used refine the model. First, the previous 9 data points are graphed with these 3 new points (**Figure 10**):

```
> with(plots):
> newA:=plot([[t[0],p[0]], [t[1],p[1]], [t[2],p[2]], [t[3],p[3]],
            [t[4],p[4]], [t[5],p[5]], [t[6],p[6]], [t[7],p[7]],
            [t[8],p[8]], [t[9],p[9]], [t[11],p[11]], [t[13],p[13]],
            [t[14],p[14]]], x=1900..2000, y=0..60000000,
            style=point, title='data points', symbol=box,
            labels=[time, population]):
> display(newA);
```

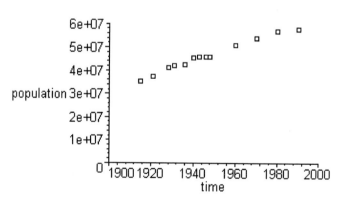

Figure 10. Plot with points for 1970, 1980, and 1990 added.

From this plot, it appears that M should be about 60 million. Estimation of M is very much tied to the history of the country being considered. For Italy, 50 million seemed like a very good estimate for M before 1950, but this changed in the ensuing decades. Using the revised M, the modeling function $L(t)$ becomes:

```
> L:=t-> M/(1+B*exp(-M*k*(t-1915))):  M:=60000000:  L(t);
```

$$L(t) = \frac{60000000}{1 + Be^{-60000000k(t-1915)}}$$

```
> t[0]:=1915:  p[0]:=35240000:  solve(p[0]=L(t[0]),B):
> B:=evalf(%);
```

$$B := .7026106697$$

```
> L(t);
```

$$L(t) = \frac{60000000}{1 + .7026106697e^{-60000000k(t-1915)}}$$

Then t_{11} and p_{11} are used to find k. This data point is used since it is the middle point of the three new points being used to update the model.

```
> solve(p[11]=L(t[11]),k);
```

$$.5407883627 \ 10^{-9}$$

```
> k:=%:
```

Thus, after simplifying, $L(t)$ looks like:

```
> L(t);
```

$$L(t) = \frac{60000000}{1 + .7026106697e^{-.03244730176t+62.13658287}}$$

The values of $L(t)$ at times t_0 through t_{14} and the actual populations through the year 1990 are:

```
> seq(evalf(L(t[i])), i=0..14);
```

$.3523999999 \ 10^8, .3801521937 \ 10^8, .4107307488 \ 10^8, .4231106310 \ 10^8,$
$.4426545184 \ 10^8, .4572499411 \ 10^8, .4675681357 \ 10^8, .4773405840 \ 10^8,$
$.4835519319 \ 10^8, .5158402926 \ 10^8, .5269086676 \ 10^8, .5367000000 \ 10^8,$
$.5453158102 \ 10^8, .5528619135 \ 10^8, .5651651603 \ 10^8$

```
> seq(evalf(p[i]), i=0..14);
```

$.35240000 \ 10^8, .37270493 \ 10^8, .41168000 \ 10^8, .42118835 \ 10^8,$
$.42527561 \ 10^8, .45330441 \ 10^8, .45801000 \ 10^8, .45646000 \ 10^8,$
$.45706000 \ 10^8, .50763000 \ 10^8, .52736000 \ 10^8, .53670000 \ 10^8,$
$.55810000 \ 10^8, .57040000 \ 10^8, .57657000 \ 10^8$

Figure 11 shows how the graph of $L(t)$ fits the data points:

```
> with(plots):
> newBC:=plot(L(t),t=1900..2000,y=0..60000000,
              labels=[time,population], color=blue,,
              title='new L(t) and data points'):
> display(newA,newBC);
```

new L(t) and data points

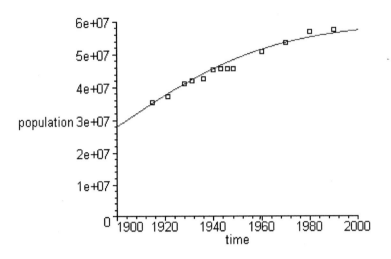

Figure 11. Data points and the improved model.

Using this new model to predict the population of Italy over the next 50 years gives:

```
> L(2000), L(2010), L(2020), L(2050);
```

$.5744058647 \ 10^8, .5812764801 \ 10^8, .5863465534 \ 10^8, .5947677938 \ 10^8$

These model predictions can then be compared to estimates obtained using other projection techniques. For example, the program `IntlPop` uses a version of the cohort method to make population projections (see the Internet sites in Section 5.2 for information about `http://geosim.cs.vt.edu/`, the home page of Project GeoSim at Virginia Tech where this program is available). `IntlPop` gives the following the population projections for Italy in the years 2000, 2010, 2020, and 2050:

$.58264000 \ 10^8, \ .57778000 \ 10^8, \ .55498000 \ 10^8, \ \text{and} \ .44718000 \ 10^8.$

There is a great discrepancy between the projections of the logistic model developed above and `IntlPop`. One of the reasons is that the model in the `IntlPop` simulations assumes that the total fertility rate in Italy will remain at the low level of 1.6 (below the replacement rate) over the next half century, contributing to an overall decline in the population. Since the logistic curve is always increasing, it is an inappropriate model for periods of population decline.

This example illustrates some of the difficulties inherent in human population modeling. Exponential and logistic models are very accurate and useful

18

(even for predicting future growth) in a controlled laboratory setting, for getting quite accurate models for backward analysis of human population change, and even for very short-term predictions. However, both models developed in this example failed as long-range predictors of Italy's future population change. The logistic model considers only relations between the overall rate of growth and the carrying capacity of the system; in particular, it could not model the effects of war or the current low fertility rate.

For example, what criteria should be used to evaluate population models? What population characteristics should be incorporated into a model? More-sophisticated models factor in such variables as gender-differentiated birth and death rates, changes due to immigration, and changes within segregated age groups. Is it even possible to develop models that reflect the possibility of war, natural disaster, medical advances, or social change?

To get an idea of some of mathematical methods used in demography (as well as biology and ecology), consult the supplementary resource list in **Section 5.3**. This includes a listing of some of the pertinent journals, early works in these fields, and reference sources for issues in these areas.

3. Modeling Activities

3.1 Examples with Given Data

The following are "warm-up" exercises in modeling with various functions. The data are provided, as well as questions to guide the analysis.

3.1.1 Exponential Growth

Suppose yeast is cultured at noon on Monday. When the culture is examined at 10:00 A.M. on Tuesday, there are 1.2 thousand organisms present. At 1:45 P.M. on Friday, there are 3.6 thousand organisms. The culture is assumed to grow exponentially.

1. Find a function $P(t)$ for the number of yeast organisms present t hours after noon on Monday (remember to convert hours into decimals).

2. How many organisms were present initially (i.e., at noon on Monday)?

3. What is the growth constant and what does this number mean?

4. How many organisms were there at noon on Wednesday?

5. When will there be 5 thousand organisms present?

6. How fast will the population be growing at that time? How could you estimate this without using the derivative?

7. How long does it take for the population to double in size? How long to quadruple? Is there a pattern here?

8. Graph $P(t)$ and use it to visually check if the answers seem accurate.

9. If another measurement were taken on Saturday, what kind of count would support using an exponential model? What count would suggest using a logistic model?

10. If another measurement were taken on Thursday, what kind of count would support using an exponential model? What count would suggest using a limited growth model?

11. What characteristics of an experiment would contribute to an exponential model being a reasonable choice?

12. What experimental factors might affect how reliable this model is as a predictor of growth after 1:45 P.M. on Friday?

3.1.2   Logistic Growth

Some yeast was grown in a petri dish, and the diameter of one colony was measured, giving the data in **Table 3**. As in the example in Section 2.3, find a logistic curve to model this data, and then answer the following questions:

Table 3.

Yeast data.

date in Sept.	time	diameter (mm)
3	10:00 A.M.	0.75
4	9:10 A.M.	1.5
5	11:55 A.M.	3
7	12:30 P.M.	4
9	2:51 P.M.	5.1
10	9:30 A.M.	5.25
11	10:10 A.M.	5.33
12	6:15 A.M.	5.39

1. Estimate the area covered by the colony at 10:25 A.M. on Sept. 6. How accurate does this estimate seem to be?

2. At what rate is the area increasing at that time? What does this mean about the growth?

3. What is the carrying capacity of the system? What does this represent physically?

20

4. When does the rate of growth slow down? In other words, when does the population stop growing at an increasing rate and start growing at a decreasing rate? Why does this happen?

5. What is the average area covered by the colony during the course of the experiment? Why might someone be interested in this?

6. Could this model be used to extrapolate to areas after Sept. 12? For about how long might the predictions be usable?

7. Geberate a graph of the data points, a graph of the logistic curve, and a graph of the logistic curve with the data points superimposed on it.

3.2   Laboratory Analysis

During the course of this class, data will be collected from several laboratory experiments. For each experiment, a table like the one below can be used to record data. The questions following the table guide the modeling and analysis.

Experiment: _____

Organism observed: _____

Measurement method: _____

Observation #	Date & Time	No. of Hours (in decimals)	Observation
⋮	⋮	⋮	⋮

1. Decide what kind of curve (e.g., exponential or logistic) seems to fit data. Justify the choice of the model.

2. Use Maple to get a specific function that models the data. The work should include the following:

    a) a graph of the data points,

    b) work showing how the growth and other constants were determined,

    c) a comparison between the actual data and the values predicted by the modeling function,

    d) a plot of the function, and

    e) a plot of the function with the actual data points plotted on the same graph.

Note: It may be necessary to try several different values for the constants (e.g., try different data points to solve for the growth constants, adjust your guess for the carrying capacity M, etc.) to get a curve that models the data well.

3. Answer the following questions. These are questions that could not have been answered just from the experimental data but that can be answered using the tools of calculus now that there is a mathematical model for the experiment. Show the calculations used to find the answers to these questions, being sure to include units and comments when appropriate.

   a) What are the constants used, and what do they represent physically?

   b) Estimate the area covered by the colony at 10:00 A.M. on the 5th day of the experiment? How accurate does this estimate seem to be?

   c) At what rate is the growth increasing at that time? What does this mean about the growth?

   d) Was the rate at 10:00 A.M. on the 5th day greater or less than the rate at 10:00 A.M. on the 8th day? What does this mean, and what aspects of the experiment might account for it?

   e) Did the rate of growth slow down or change in any way during the course of the experiment? If so, when did the change occur? What might have caused this?

   f) What is the average area covered by the colony during the course of the experiment?

   g) What would the population be 24 hours after the last observation? What about 48 hours later? For approximately how long might predictions from this model be usable? Why?

   h) Comment on your results. How well do they seem to reflect or predict reality?

3.3   Human Populations

After the success in modeling the growth observed in biological experiments, it seems reasonable to use similar modeling techniques to predict human population levels based on current demographic data. This involves the following:

- Creation of human population models for various countries based on recent population levels (to test model accuracy, earlier population data can be used to project figures for years in which census data is available).

- Use of library and Internet resources to gather demographic information (see Sections 5.2 and 5.3 for some useful sources to get started).

- Application of the population simulation program Intlpop, which provides such data as recent population levels, life expectancies, fertility rates, infant mortality rates, and net migration data for various countries/regions in the world. Other sites listed in Section 5.2 provide similar population data for various regions in the world and in the United States.

- Discussion of the politics of census figures and population projections, leading to the realization that different sources often provide very different projections (Haub [1987] is helpful).

3.3.1 Analyzing the Population of a Specific Country

The following activities use the concepts developed in this Module to study the original question of human population change.

- Choose a country of interest. Using library or Internet resources, collect as much population data as possible prior to 1950.

- With these data, use Maple to develop a model for the population of that country. Explain the choice of modeling function.

- Use the modeling function to predict the population of that country in 1960, 1970, and 1980.

- Collect data on the actual population of the country in 1960, 1970, and 1980. What might account for any discrepancies? This may require some research into the history of the country.

- Use the data from the 1980s and 1990s to refine the modeling function. If you change the modeling function, explain how and why.

- Compare the actual data with the populations given by the modeling function and seek reasons for any discrepancies. How well does the model estimate the actual populations in 1965 and 1975? Compare function values to actual data if available.

- Use the new modeling function to predict the population of the country for the years 2000, 2010, and 2020.

- Research population literature (including Intlpop) to get predictions for the future population of the country. How does the model's predictions compare with those in the literature? What might account for any discrepancies? ("They have a better model" is not an adequate answer—it is important to discuss what factors their model uses that your model does not and to consider differences in the assumptions made.)

- How are human populations different from organisms grown in the lab? How might the modeling functions be modified to give better predictions?

- As an exercise in examining how models can be manipulated to achieve a desired result from political or ideological motives, or even to pander to Hollywood sensationalism, analyze the premise of the movie *Soylent Green*. First, find the population of New York City in 1973 when the movie was made. Then develop a population model supporting the movie's premise that the population of NYC in 2022 will be 40 million. What does this model predict that the population of NYC would have to be today? Find the most recent census data available for the actual population of NYC today and compare it to the value predicted by the model. Comment on the results.

4.   Developing Facility with Maple

4.1   Examples and Discussion

The following examples provide practice using Maple and give interpretations of various mathematical concepts in the physical context of modeling population changes. The examples are given in code for Maple V, Release 5. Most of this code should still work with earlier releases, noting that % has replaced " in Release 5. These examples and the following exercises could be modified for use with other computer algebra systems such as Mathematica or Derive, or even with a good graphing calculator.

4.1.1   Graphing

A common use for Maple in this Module is to define a function and plot its graph. Limits for the domain values must be given, but the limits for the range values are optional, as in

```
> f := x-> (x-3)^2 + 4*x - 10;
> plot(f(x),x=-5..5,y=-5..5);
```

To "zoom in" on the graph, simply restrict the x and y ranges, as in

```
> plot(f(x),x=2.3..2.5,y=-0.05..0.05);
```

At this magnification, the curve looks almost linear, which indicates that locally a linear equation (the equation of the tangent line, in fact) gives a good approximation of the function.

Maple can also plot individual data points, as follows:

```
> plot ( [ [1,2], [-3,1], [1.5, 3], [-1,-1], [2,0.5] ], x= -4..3,
         y=-2..4, style=point, symbol=diamond );
```

4.1.2 Solving

Maple can be used to solve equations and systems of equations and to find roots of functions. For example, to find the roots of $f(x) = (x - 3)^2 + 4x - 10$, use the solve command:

```
> solve(f(x)=0,x);
```

This command can be used to determine the time of an event by finding solutions to a given equation. For example, suppose that the concentration of a drug in the blood stream t hours after noon is given by $A(t) = 1.36e^{-0.34t}$. The time at which the concentration of the drug is 0.5 milligrams can be found by:

```
> solve(0.5=1.36*exp(-.34*t));
```

Maple can solve also systems of equations. To illustrate this, suppose that a yeast culture is growing exponentially according to $p(t) = Ce^{kt}$. Assume that 2 days after the experiment begins, there are 140 thousand organisms present; and 3.5 days after, there are 300 thousand present. To find an exponential function $p(t)$ modeling the yeast growth, use

```
> fsolve({140=C*exp(k*2), 300=C*exp(k*3.5)}, {C,k});
```

The difference between the commands solve and fsolve is that solve will return exact solutions, such as $\sqrt{2}$, if possible, while fsolve returns floating-point decimals.

4.1.3 Differentiating

Maple differentiates functions using the diff command. The first argument is the function and the second is the independent variable. For example, differentiate $V(r) = \pi r^2 h$ with respect to r:

```
> diff(Pi*r^2*h, r);
```

The diff command can also be used to show that a function is a solution to a differential equation. For example, here's how to show that $y(t) = Ce^{kt}$ is a solution to $y' = ky$:

```
> y:=t->C*exp(k*t):
> simplify (diff(y(t), t)-k*y(t));
```

The result of 0 shows that $y' - ky = 0$ and hence $y' = ky$.

4.1.4   The Shapes of Graphs

The following functions have plots that exhibit the four basic shapes that may occur in a graph.

```
> f1:=x->exp(x):   f2:=x->ln(x):   f3:=x->-ln(x):   f4:=x->20-exp(x):
```

● > `plot(f1(x), x=-3..3);`

This function is increasing at an increasing rate (increasing and concave up). It might be the shape of a function modeling population growing at an increasing rate, for example bacteria growing with unlimited nutrients.

● > `plot(f2(x), x=-3..3);`

This function is increasing at a decreasing rate (increasing and concave down). It might represent a population with some limiting factor.

● > `plot(f3(x), x=-1..3);`

This function is decreasing at an increasing rate (decreasing and concave up), representing a gradually declining population, perhaps one where the birth rate is slightly less than the mortality rate.

● > `plot(f4(x), x=-3..3);`

This function is decreasing at a decreasing rate (decreasing and concave down). It could model a population experiencing a precipitous drop off in numbers, perhaps following disease or disaster.

4.1.5   Exponential and Logarithmic Functions

Exponential and logarithmic functions are essential to the models developed in this module. An exponential function has the form $y = a^k$, for some fixed constant a. Here are some examples of exponential functions:

```
> plot({(1/2)^x, 2^x}, x=-5..5);
> plot({(1/3)^x, 3^x}, x=-3..3);
> plot({seq(i^x, i=2..5)}, x=-2..2);
> plot({2^x, exp(x), 3^x}, x=-3..3);
```

Notice that $y = a^k$ and $y = a^{-k}$ are symmetric about the y-axis and that $y = 0$ is a horizontal asymptote for all such functions. Also, the graph of e^x lies between the graphs of 2^x and 3^x, just as e lies between 2 and 3.

Now consider the graphs of some logarithmic functions. The Maple notation for $\log_b a$ is log[b](a). Recall that $\log_b a = c$ means that $b^c = a$. Also, the natural logarithm is $\ln x = \log_e x$.

```
> plot({log[1/2](x), log[2](x)}, x=-1..100);
> plot({log[2](x), ln(x), log[3](x), log[4](x), log[10](x), 1},
         x=0..10);
```

Notice that $y = \log_a x$ and $y = \log_{a^{-1}} x$ are symmetric about the x-axis and that $y = \log_a x$ intersects the line $y = 1$ at the point $(a, 1)$.

The following graphs illustrate the inverse relationship between $y = a^x$ and $y = \log_a x$ and show that they are reflections of each other about the line $y = x$.

```
> plot({x, 2^x, log[2](x)}, x=-5..5, y=-5..5);
> plot({x, exp(x), ln(x)}, x=-5..5, y=-5..5);
```

Plotting a few exponential functions with their derivatives shows that the derivative of $y = a^x$ (which turns out to be $y = a^x \ln a$) is very similar to the exponential function itself, and when $a = e$ the function and its derivative are identical.

```
> f:=(a,x)->a^x:
> g:=(a,x)->diff(f(a,x),x):
> plot({f(2,x),g(2,x)}, x=-3..3,y=0..8,
          title='2^x and its derivative');
> plot({f(3,x),g(3,x)}, x=-3..3, y=0..8,
          title='3^x and its derivative');
> plot({f(exp(1),x),g(exp(1),x)}, x=-3..3, y=0..8,
          title='exp(x) and its derivative' );
```

4.2 Maple Exercises

The following problems provide practice using Maple to perform the types of calculations that are likely to arise in this Module.

1. Graphing and zooming

   a) Graph $f(x) = e^x$.

   b) Graph $g(x) = Ce^{kx}$, where $C = 0.0386$ and $k = 2.18$.

   c) Graph $h(t) = \dfrac{M}{1 + Be^{-kMt}}$, where $M = 32$, $B = 21.5$, and $k = 0.18$.

   d) Graph $j(x) = 2x^2 - 15x - 10$ from $x = -5$ to 10 and estimate the zeros of $j(x)$ by zooming in.

   e) Plot the time and number of people in the computer lab for four different times. Measure times starting with the first observation and converting them from hours/minutes to decimals. Also, use reasonable x and y ranges to display the data. For example, suppose observations at 12:30, 1:00, 1:15, and 1:50 find that there are 10, 8, 3, and 18 people in the lab. The data points would then be [0, 10], [0.5, 8], [0.75, 3], and [1.33,18]. Reasonable x and y ranges for this example would be $x = 0 \ldots 1.5$, $y = 0 \ldots 20$.

2. Solving

    a) Solve for x: $x^2 - 4x - 6 = 0$.

    b) Solve for x: $ax^2 + bx + c = 0$.

    c) Find the roots of $h(x) = x^3 - 3x$ by setting it equal to zero and then solving.

    d) If $f(t) = 2.30e^{-0.2t}$ is the concentration of a drug in the blood at time t, find when the concentration will be 2.9 milligrams by finding the roots of $f(t) - 2.9$. Then determine when the concentration of the drug will be 1.3 milligrams, and give possible interpretations of the solution.

    e) Solve the system of equations: $y = 3x - 5$, $3y - 2x = 12$.

    f) Suppose a bacterial culture is growing exponentially, 200 thousand bacteria are present 1 hour after the experiment begins, and 265 thousand are present after 2.5 hours. Find the constants C and k by solving $p(1) = 200$ and $p(2.5) = 265$, where $p(t) = Ce^{kt}$.

    g) Unfortunately, when very large or small numbers are involved, Maple (or any other CAS or a graphing calculator) may give an error message. For example, if solving $p(1) = 200$ and $p(26) = 950$ for C and k, where $p(t) = Ce^{kt}$, yields an error message using the technique above, it is still possible to solve for k by using the fact that for exponential functions the growth constant is equal to

$$\frac{\ln A_1 - \ln A_2}{t_1 - t_2},$$

    where A_1 is the amount at time t_1 and A_2 is the amount at time t_2. Thus, it is possible to find k first and then solve for C. Carry out this computation.

3. Differentiation

    a) Differentiate $f(x) = x^3 - x^2$.

    b) Differentiate $g(t) = Ce^{kt}$ with respect to t.

4. Differential Equations

    a) Show that $y = Ce^{kt}$ satisfies the differential equation $y' = ky$.

    b) Show that $v(t) = M\left(1 - e^{-kt}\right)$ is a solution to the differential equation $v' = k(M - v)$.

    c) Show that $y(t) = M/\left(1 + Be^{-Mkt}\right)$ satisfies the differential equation $y' = ky(M - y)$.

5. Shapes of Graphs

Plot the graphs of the functions below, adjusting the y-range as needed to get a good picture. Although these graphs are unlikely to represent models of population growth, consider what their shapes would indicate about the way the population would be changing. What might the concavity, asymptotes, inflection points, and relative extrema indicate about the behavior of a population?

a) $f(x) = 10x(x-0.5)(x+1.5)(x-1)(x+0.5)(x+0.75)$, for $-1.75 \leq x \leq 1.5$.

b) $f(x) = \dfrac{x^2 - 4}{(x+1)(x-3)}$, for $-6 \leq x \leq 6$.

c) $f(x) = x^3 - x^2$, for $-1 \leq x \leq 1$.

5. Resources

5.1 Laboratory Notes

Below is the basic information for each lab that can be included in this Module. For each experiment, the necessary equipment is listed and the procedure is specified. If there is not time or facilities to do all of the experiments, it suffices to choose one example of exponential growth and one of logistic growth.

The populations studied include slime molds (to model qualitatively exponential growth), flour beetles (to model exponential growth), yeast cultures (to model both exponential and logistic growth), and ciliated protistans (to model exponential and logistic growth).

Some materials can be obtained at a grocery store. Other items can be ordered from biological supply companies, such as the following:

Carolina Biological Supply Company
2700 York Road
Burlington, NC 27215
(800) 334–5551

Connecticut Valley Biological
82 Valley Road
P.O. Box 326
Southampton, MA 01073
(413) 527–4030

5.1.1 Exponential Growth with Slime Molds and Flour Beetles

Equipment needed (for slime mold experiment)
Slime molds (e.g., *Physarum polycephalum* or *Dictyostelium discoideum*)
Non-nutrient agar (2% agar)

Oatmeal
Petri dishes

Slime Mold Procedure:
The slime molds are grown in a petri dish on sterile agar with oatmeal sprinkled on top. When the slime molds run out of oatmeal, they will move out in search of food. The actual timing of this "migration" depends on the initial amount of food (this could even become the basis of an experiment, if so desired). It will most likely take at least 2 weeks for the slime molds to begin migrating.

Flour Beetle Information:
The experiment with flour beetles is outlined in Part B of Glase and Zimmerman [1993]. Further descriptions of this experiment and the population changes being modeled can be found in Glase and Zimmerman [1992]. The beetles provide a good example of exponential growth, but it takes upwards of 7 weeks to collect sufficient data.

5.1.2   Exponential/Logistic Growth with a Yeast Culture

A yeast culture exhibits exponential growth during the first 10 to 36 hours of the experiment and logistic growth if the experiment continues for 48 hours.

Equipment needed
Culture tubes containing 15 ml of YPD broth (consisting of 1% yeast extract,
    2% peptone, and 2% dextrose)
Counting chamber (a spectrophotometer can also be used)
Microscope (400X)
Yeast culture (dry yeast available in grocery stores can be used)
Orbital shaker

Yeast Procedure:
Add one grain of dried yeast to 15 ml of sterile YPD in a culture tube. Gently dissolve the yeast grain. Incubate the tube at room temperature, or for faster growth at 30°C. Periodically (and at least at the beginning of the experiment), remove 10 μl of broth and place on the counting chamber. At room temperature, observations should be made every 4 hours or so. Count the number of cells to determine the concentration of yeast cells. The small squares of the counting chamber have an area of 1/400 mm^2. The depth of the liquid is 1/50 mm. Cell concentration is reported as cells/ml. If a spectrophotometer is used, then the absorbance of the culture can be measured at 600 nm.

5.1.3   Logistic Growth with a Bacteria Culture

This is perhaps the least complicated of all the experiments. In it, bacterial cultures get restricted nutrients and exhibit logistic growth. The amount of time needed to see the logistic growth depends on the incubation temperature of the agar plates. At 35°C, it could take 24 to 36 hours, whereas at room temperature it may require several days.

Equipment needed
Bacteria culture (*Bacillus subtilis*)
Tryptic soy agar plate (DIFCO)
Ruler
Magnifying glass

Bacteria Procedure:
Streak out *Bacillus subtilis* on a tryptic soy agar plate (DIFCO) to obtain well-separated colonies. To measure colony growth accurately, it may be most effective to isolate a single culture on a single plate. The plates can be incubated at either room temperature or 35°C, depending on the desired rate of growth. At room temperature, observations should be made about once a day. The diameters of the resulting bacterial colonies can be measured with a ruler and magnifying glass.

5.1.4 Population Crash with Ciliated Protistans

This experiment demonstrates exponential and logistic growth followed by a precipitous decline or "crash" in population using a few different species of ciliated protistans.

Equipment needed
Handout entitled "Population Ecology: Experiments with Protistans"
Protistan cultures
Stereoscopic binocular microscope
Pasteur pipettes and bulbs
Volumetric pipettes and dispensers
Depression slides and counting plates or hemocytometer
Sterile spring water
Culture vials and plugs
Concentrated liquid food

Protistan Procedure:
The experimental procedure is outlined in detail in Part A of "Population Ecology" (see **References**). Four to five different protistan species are available for selection. One type containing photosynthetic green algae (*Paramecium bursarie*) is useful in experiments of light vs. dark growth. For the purposes of this Module, it may be most worthwhile to simplify the parameters of the experiment as much as possible and to set up only single species cultures. The "Population Ecology" handout discusses more complicated variations of a single species study, such as predator/prey and competition experiments, which may be worthwhile if time permits. Protistan cultures are grown in culture vials, and pp. 42–43 of "Population Ecology" gives precise and detailed techniques

for counting individual organisms. For example, after thoroughly mixing the vials, Pasteur pipettes are used to place sample of the cultures onto counting plates for microscopic inspection. Using these methods, protistan growth can be closely monitored and then modeled.

5.2   Population Internet Sites

http://www.census.gov/
Home page of the United States Census Bureau
This site offers a rich collection of social, demographic, and economic information. For example, one can search for population information indexed by word or by place, or on a clickable map. Using a cohort-component model, the Census Bureau offers population projections from 1996 to 2050 for United States resident population based on age, race, sex, and Hispanic origin. This site also has population clocks for the United States and world as well as information about Census 2000 in the United States.

http://geosim.cs.vt.edu/
Home page of Project GeoSim at Virginia Tech
Project GeoSim offers many educational modules for introductory geography courses, including HumPop and IntlPop. HumPop is a multimedia tutorial program introducing and illustrating population concepts and issues, while IntlPop is a useful population simulation program for various countries and regions around the world.

http://www.nidi.nl/links/nidi6000.html
Home page of NiDi
Maintained by the Netherlands Interdisciplinary Demographic Institute, this site offers a comprehensive overview of demographic resources on the Internet. It contains about 400 external links to various useful sites about software and demographic models, census/survey and data facilities, research institutes and organizations, literature, conferences, and other information resources.

http://www.popnet.org/
Home page of Popnet
Created by the Population Reference Bureau, Popnet offers links to many sites containing global population information. For example, it contains links to organizational sources in many countries throughout the world, offers various demographic statistics, and includes a "demographic news update." It also contains a clickable world map through which the user can obtain a directory of websites that provide region specific information.

http://www.prb.org/
Home page of the Population Reference Bureau
This site contains useful information on United States and international population trends. For example, it offers the World Population Data Sheet, an annual

publication containing recent population estimates, projections, and key indicators for areas with populations of 150,000 or more as well as all members of the United Nations. This site also links to reports on a variety of population issues.

http://www.iisd.ca/linkages/
Home page of Linkages
Provided by the International Institute for Sustainable Development, Linkages is a clearing-house for information on previous and upcoming international meetings related to development and environment. For example, it contains a link (http://www.iisd.ca/linkages/cairo.html) to the home page of the 1994 Cairo International Conference on Population and Development as well as links to current conferences on population.

http://popindex.princeton.edu/
Home page of the Population Index
This site contains an online version of the Population Index, a primary reference tool offering an annotated bibliography of recently published journal articles, books, working papers, and other materials on population topics. The site has an online database (for 1986–1999) which can be searched by author, geographical region, subject matter, and year of publication.

http://coombs.anu.edu.au/ResFacilities/DemographyPage.html
Home page of Demography and Population Studies at ANU
Provided by Australian National University, this regularly updated site has an assortment of links to demographic information centers and facilities worldwide. These include links to various demography and population studies WWW servers, databases of interest to demographers, and demography and population studies gopher servers.

http://www.popassoc.org/
Home page of the Population Association of America
This site offers information about the Population Association of America and its quarterly journal Demography. Articles from this journal can be accessed using the full text archive JSTOR (which is linked to this site at http://www.jstor.org/jstor/), through which articles from other pertinent journals also can be obtained.

5.3 Additional Resources

5.3.1 Early Works

Gause, G.F. 1934. *The Struggle for Existence*. Baltimore, MD: Williams and Wilkins.

Lotka, A.J. 1925. *Elements of Physical Biology*. Baltimore, MD: Williams and Wilkins. 1956. Reprinted as *Elements of Mathematical Biology*. New York: Dover.

Malthus, T.R. 1798. *An Essay on the Principle of Population As It Affects the Future Improvement of Society*. London, England: J. Johnson.

Pearl, Raymond, and L.J. Reed. 1920. On the rate of growth of the population of the United States since 1790 and its mathematical representation. *Proceedings of the National Academy of Science* 6: 275–288.

Verhulst, P.F. 1938. Notice sur la loi que la population suit dans son accroissement [Note on the law followed by a growing population]. *Correspondance Mathématique et Physique* (Brussels) 10: 113–121.

Volterra, Vito. 1926. Variations and fluctuations in the number of individuals of animal species living together. Appendix in *Animal Ecology*, edited by R.N. Chapman, 409–448. New York: McGraw-Hill.

5.3.2   Biology

Krebs, C.J. 1972. *Ecology: The Experimental Analysis of Distribution and Abundance*. New York: Harper & Row. A good review of classical population experiments of the first half of the twentieth century on yeast, bacteria, *Drosophila*, and flour beetles is provided on pp. 190–200.

May, R.M., ed. 1976. *Theoretical Ecology Principles and Applications*. Philadelphia, PA: W.B. Saunders.

Seeley, H.W., and P.J. VanDemark. 1981. *Microbes in Action*, 82–83. San Francisco, CA: W.H. Freeman.

5.3.3   Demography

Keyfitz, Nathan. 1968. *Introduction to the Mathematics of Population*. Reading, MA: Addison-Wesley. 1977. 2nd ed.

_____. 1976. Mathematical demography: A bibliographical essay. *Population Index* 42 (1) (January 1976): 9–38.

_____. 1977. *Applied Mathematical Demography*. New York: Wiley. 1985. 2nd ed.

Shryock, H.S., J.S. Siegel, and associates. 1973. *The Methods and Materials of Demography*, vol. 2. Washington, DC: U.S. Bureau of the Census.

Wattenberg, Ben. 1991. *The First Universal Nation*. New York: Free Press.

The World Almanac and Book of Facts. 1916 and 1924 eds. New York: New York World. 1931, 1933, 1938, 1942, 1946, 1948, 1950 eds. New York: New York World-Telegram 1961 ed. New York: New York World-Telegram & The Sun. 1967, 1972, 1977, 1982 eds. New York: Newspaper Enterprise Association. 1991 ed. New York: Pharos Books.

5.3.4 Curve Fitting and the Logistic Model

Cavallini, Fabio. 1993. Fitting a logistic curve to data. *College Mathematics Journal* 24 (3) (May 1993): 247–253.

Macchetti, C., P. Meyer, P., and J. Ausubel. 1996. Human population dynamics revisited with the logistic model: How much can be modeled and predicted? *Technological Forecasting and Social Change* 52 (1) (1 May 1996): 1-30.

Maruszewski, R. F. 1994. Approximating the parameters for the logistic model. *Mathematics and Computer Education* 28 (1) (Winter 1994): 16–19.

Matthews, J.H. 1992. Bounded population growth: A curve fitting lesson. *Mathematics and Computer Education* 26 (2) (Spring 1992): 169–176.

5.3.5 A Few Pertinent Journals

Demography

Ecology

Journal of the American Statistical Association

Journal of Ecology

Population and Development Review

Population Index

Population Studies

6.   Teaching Notes

Ideally, this Module would run throughout a first-semester calculus course. Some lab experiments take several weeks to develop growth patterns, and these should be started at the beginning of the course. Others take as little as a few days to complete. Meaningful discussion of the data from these labs can begin as soon as students can graph data points and functions. More in-depth analysis can continue after the definition of the derivative has been developed and various properties and examples of derivatives have been considered. The Module is designed to give a context for and hands-on experience with central concepts such as derivatives, exponential and logarithmic functions, properties of graphs, and integration in roughly the order that they are introduced in a standard elementary calculus course.

The following activities should be done early in the course, both to introduce the module and to leave enough time to collect data:

- Distribution of scenes from the science fiction movie *Soylent Green* to begin discussion of motivating sociological and political issues surrounding population changes.

- Assignment of background reading, perhaps containing dramatically different population projections, to illustrate some of the difficulties in forming accurate demographic predictions (see Alonso and Starr [1987], for example).

- Initialization of the experiments. Some of the biology experiments can take up to several weeks to run. It is important to set them up at the beginning so that students can collect data for their final projects throughout the course.

- Introduction to the Maple CAS. To do the exercises associated with these projects, students need to be able to use Maple to perform basic equation and function manipulations. In particular, they must be able to solve equations and systems of equations and to find the roots of functions. They must be able to differentiate and to verify solutions of differential equations. They must also be able to plot points and graph functions. Section 4, **Developing Facility with Maple**, is designed to help students to develop these skills.

- Practice modeling using the examples with given data in Section 3.1.

A version of the projects in this Module that is not Maple-specific and includes classroom demonstrations and student handouts is available [Ashline and Ellis-Monaghan 1999]. There is also an expository paper describing the authors' experiences with these projects [Ashline and Ellis-Monaghan 1999].

References

Alonso, William, and Paul Starr, eds. 1987. *The Politics of Numbers*. New York: Russell Sage Foundation.

Ashline, George, and Joanna Ellis-Monaghan. 1999. How many people are in your future? Elementary models of population growth. In *Making Meaning: Integrating Science Through the Case Study Approach to Teaching and Learning*, 42–80. New York: McGraw-Hill Primis.

_____. 1999. Interdisciplinary population projects in a first semester calculus course. *PRIMUS: Problems, Resources, and Issues in Mathematics Undergraduate Studies* 9 (1) (March 1999): 39–55.

Cohen, Joel. 1995. *How Many People Can the Earth Support?* New York: W.W. Norton.

Ehrlich, Paul. 1968. *The Population Bomb*. New York: Ballantine Books.

Fleisher, Richard, director. 1973. *Soylent Green*. With Charleton Heston, Leigh Taylor-Young, Chuck Connors, Joseph Cotton, Edward G. Robinson. MGM. Available on videocassette.

Glase, J.C., and Melvin Zimmerman. 1992. Studies in protozoan population ecology. In *Tested Studies for Laboratory Teaching, Proceedings of the Thirteenth Workshop/Conference of the Association for Biology Laboratory Education (ABLE)*, 19–66.

_____. 1993. Population ecology: Experiments with protistans. In *Experiments to Teach Ecology*, 37–69. Tempe, AZ: Ecological Society of America.

Harrison, Harry. 1966. *Make Room! Make Room!* Boston, MA: Gregg Press.

Haub, Carl. 1987. Understanding population projections. *Population Bulletin* 42 (4) (December 1987): 3–43.

Population Reference Bureau. 1999. *1999 World Population Data Sheet*.

Simon, Julian. 1990. *Population Matters*. New Brunswick, NJ: Transaction Publishers.

Tapinos, Georges, and Phyllis T. Piotrow. 1980. *Six Billion People*. New York: McGraw-Hill.

Wattenberg, Ben. 1987. *The Birth Dearth*. New York: Ballantine Books.

Acknowledgments

We would like to thank the following St. Michael's College professors for so enthusiastically sharing the resources of their respective fields with us: Glenn Bauer and Peter Hope (Biology), Vincent Bolduc (Sociology), and Richard Kujawa (Geography).

About the Authors

George Ashline received his B.S. from St. Lawrence University, his M.S. from the University of Notre Dame, and his Ph.D. from the University of Notre Dame in 1994 in value distribution theory. He has taught at St. Michael's College since 1995. He is a participant in Project NExT, a program created for new or recent Ph.D.s in the mathematical sciences who are interested in improving the teaching and learning of undergraduate mathematics.

Joanna Ellis-Monaghan received her B.A. from Bennington College, her M.S. from the University of Vermont, and her Ph.D. from the University of North Carolina at Chapel Hill in 1995 in algebraic combinatorics. She has taught at Bennington College, at the University of Vermont, and since 1992 at St. Michael's College. She is a proponent of active learning and has developed materials, projects, and activities to augment a variety of courses.

UMAP

Modules in
Undergraduate
Mathematics
and Its
Applications

Published in
cooperation with

The Society for
Industrial and
Applied Mathematics,

The Mathematical
Association of America,

The National Council
of Teachers of
Mathematics,

The American
Mathematical
Association of
Two-Year Colleges,

The Institute for
Operations Research
and the Management
Sciences, and

The American
Statistical Association.

Module 779

Multiple Reduction Copy Machines and Fractals

Frederick Solomon

Applications of Precalculus, Calculus and Analysis to Mathematics

COMAP, Inc., Suite 210, 57 Bedford Street, Lexington, MA 02420 (781) 862–7878

INTERMODULAR DESCRIPTION SHEET: UMAP Unit 779

TITLE: Multiple Reduction Copy Machines and Fractals

AUTHOR: Frederick Solomon
Dept. of Mathematics
Warren Wilson College
Asheville, NC 28815–9000
fsolomon@owl.warren-wilson.edu

MATHEMATICAL FIELD: Precalculus, calculus, and analysis

APPLICATION FIELD: Mathematics

TARGET AUDIENCE: Students in liberal arts mathematics, precalculus, or linear algebra.

ABSTRACT: This Module uses the idea of the Multiple Reduction Copy Machine (MRCM) to explain iterated function systems and fractals. The purpose is not to develop the theory completely but to explain enough in order to create fractals on the computer. The Module describes computer animation that is easy to use but requires some basic understanding of underlying theory. The computer animation is a Java applet that can be accessed on the Internet with a Java-enabled browser at http://www.warren-wilson.edu/~fsolomon/ZZFract.html . The animation allows graphical input of the transformations defining the fractal—the lenses, as they are called in the terminology of the Multiple Reduction Copy Machine. Using the animation while reading this Module will provide much insight into the nature of self-similarity—a central defining characterization of fractals.

PREREQUISITES: None.

RELATED UNITS: Unit 716: *Newton's Method and Fractal Patterns,* by Philip D. Straffin. *The UMAP Journal* 12 (2) (1991) 143–164. Reprinted in *Tools for Teaching 1991,* 103–124. Lexington, MA: COMAP, 1992. Reprinted in *The UMAP Library.* CD-ROM. Lexington, MA: COMAP, 1996. Available online at http://www.comap.org .

Tools for Teaching 1999, 81–107. Reprinted from *The UMAP Journal* 20 (4) (1999) 431–457.
©Copyright 1999, 2000 by COMAP, Inc. All rights reserved.

COMAP, Inc., Suite 210, 57 Bedford Street, Lexington, MA 02420
(800) 77-COMAP = (800) 772-6627, or (781) 862-7878; http://www.comap.com

Multiple Reduction Copy Machines and Fractals

Frederick Solomon
Dept. of Mathematics
Warren Wilson College
Asheville, NC 28815–9000
`fsolomon@owl.warren-wilson.edu`

Table of Contents

MODULES AND MONOGRAPHS IN UNDERGRADUATE
MATHEMATICS AND ITS APPLICATIONS (UMAP) PROJECT

The goal of UMAP is to develop, through a community of users and developers, a system of instructional modules in undergraduate mathematics and its applications, to be used to supplement existing courses and from which complete courses may eventually be built.

The Project was guided by a National Advisory Board of mathematicians, scientists, and educators. UMAP was funded by a grant from the National Science Foundation and now is supported by the Consortium for Mathematics and Its Applications (COMAP), Inc., a nonprofit corporation engaged in research and development in mathematics education.

Paul J. Campbell   Editor
Solomon Garfunkel   Executive Director, COMAP

1. Introduction

This Module uses the idea of the Multiple Reduction Copy Machine (MRCM) to explain iterated function systems and fractals. The development and nomenclature are taken from Peitgen et al. [1992]. The purpose of this Module is not to develop the theory completely but to explain enough in order to create fractals on the computer. We present a computer animation that is easy to use but requires some basic understanding of underlying theory. The computer animation is a Java applet that can be accessed on the Internet with a Java-enabled browser at:

> http://www.warren-wilson.edu/~fsolomon/ZZFract.html

The animation allows graphical input of the transformations defining the fractal—the lenses, as they are called in the terminology of the Multiple Reduction Copy Machine. Using the animation while reading this module will provide much insight into the nature of self-similarity—a central defining characterization of fractals. No specific prerequisites are needed to read this Module.

2. Definitions of Terms and Some Examples

A Multiple Reduction Copy Machine (MRCM for short) is a logical machine: It exists only in the mind. The MRCM is like a copying machine but with more capabilities. An MRCM consists of one or more lenses. An input image is fed into a lens, just as an image is fed into a copier machine. But the lens can shrink, rotate, distort, and reposition the input image. The MRCM takes the resulting images—one from each lens—and overlays them to assemble a final composite output image.

The above explanation is abstract; the best way to understand how an MRCM works is through an example. The large square together with the three smaller squares in **Figure 1** conveniently describe the MRCM called the *Sierpinski Triangle MRCM*, the best-known MRCM. The three smaller squares—each with the letter L in the upper left—are the lenses.

Here is how it works: Start with any shape whatsoever. To be precise, we start with the letter F centered in the square in **Figure 2**. Copy the F three times—once for each of the smaller squares, the lenses. But reduce the size by one-half, since each lens has dimensions that are half the dimensions of the large square. That is, each lens corresponds to a copy machine with a reduction factor of 50%. We now have three new half-size letter Fs. The final step is to place these smaller Fs centered in the smaller squares.

Figure 2 shows the original image—the letter F. **Figure 3** shows the output from one run through the MRCM. Technically speaking, **Figure 3** depicts the

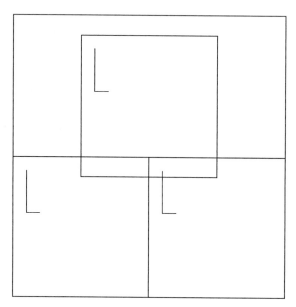

Figure 1. Sierpinski Triangle MRCM.

first iteration and **Figure 2** depicts the zeroth iteration. We iterate a second time; that is, we take the image in **Figure 3**, copy it three times—but reduce it in size by one-half—and place each of the copies centered in each of the three lenses. The result is **Figure 4a**—the second iteration. **Figure 4** also shows the images resulting from the third through the eighth iterations.

Imagine performing this iteration process over and over—infinitely many times. Will a stable shape emerge?

Figure 2. Zeroth iteration.

Figure 3. First iteration.

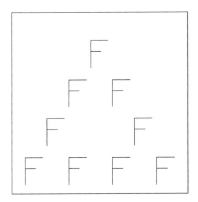

Figure 4a. Second iteration.

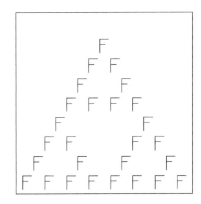

Figure 4b. Third iteration.

Figure 4c. Fourth iteration.

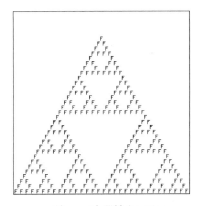

Figure 4d. Fifth iteration.

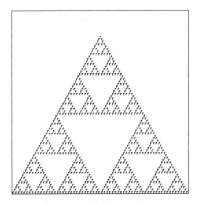

Figure 4e. Sixth iteration.

3

Figure 4f. Seventh iteration.

Figure 4g. Eighth iteration.

You can see that the seventh and eighth iterations produce identical images as far as we can tell. That a stable image emerges after many iterations seems a reasonable conclusion. In our current example, the final image is called the *Sierpinski Triangle*. In fact, relative to the finite resolution of the computer image from which the eighth iteration was printed, this iteration really is indistinguishable from the actual Sierpinski Triangle—the image emerging after infinitely many iterations, that is, in the limit as the number of iterations increases without bound. Actually, an image on paper or the computer monitor can only approximate the Sierpinski Triangle, since the real triangle has infinitely many holes.

The MRCM for the Sierpinski Triangle is the set of three lenses together with the larger square in **Figure 1**. The Ls in the upper left of each lens are there to show the orientation of the lens. In other examples, we will see that a lens can rotate and distort the original image. The L will be necessary to see this. An *iterated function system* is another name for an MRCM. Or, to be more precise, the iterated function system is the set of operations corresponding to the MRCM; each operation corresponds to one of the lenses. When we use the term "MRCM" rather than "iterated function system," we are thinking more visually. Thus, the Sierpinski Triangle MRCM is the large square together with the lenses depicted in **Figure 1**. The final image is called the *attractor*, the *attracting fractal*, or simply the *fractal* produced by the iterated function system or MRCM.

Note that the letter F is too small to be resolved in the seventh and eighth iterations in **Figure 4**. In fact, beginning with any image whatsoever—the letter F, or a portrait of Bugs Bunny, or any finite image—the result of infinitely many iterations is an image that is completely independent of the initial image. The final image is the attractor or fractal for the MRCM. The only requirement is that the initial image must be finite: It must be contained in a finite area of the plane. That the final image is independent of the initial image is a deep theorem that we will not prove here. The goal of this Module is not to prove results but simply to develop nomenclature and provide understanding to use the Java applet. You can find a proof in Peitgen et al. [1992].

Here is a summary of the key ideas:

- An MRCM is a square together with a finite number of lenses, each of which has area less than the area of the square. Each lens is a parallelogram (including square and rectangle as possibilities). The orientation of the parallelogram is determined by the letter L.

- The lenses are restricted in size to have area less than the area of the square. Otherwise the image which is copied and reproduced once for each lens might grow larger and larger rather than be restricted to a finite area.

- The fundamental MRCM result is the following theorem:

  > *Given an MRCM, regardless of what finite image is used initially, eventually, as n becomes very large, the nth iterate will more and more closely resemble an image which is called the attractor for the MRCM. The attractor is a fractal: It is composed of identical copies of itself, one in each lens.*

The number n of iterations required so that the nth iteration is visually indistinguishable from the actual attractor varies depending on the sizes of the lenses. If the area of each lens is no more than half the area of the large square, eight iterations are usually enough to approximate the actual attractor.

Figure 5 shows the Sierpinski Triangle together with the MRCM overlaid. Note this surprising feature of **Figure 5**: An exact replica of the whole image is contained in each of the lenses. This is a feature that is true of all the fractals in this Module: The fractal is a shape that is composed of smaller copies of itself.

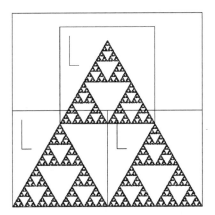

Figure 5. Sierpinski Triangle together with the MRCM overlaid.

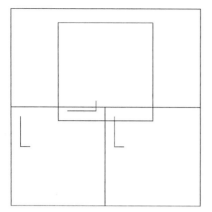

Figure 6. The Sierpinski Triangle MRCM except that the upper lens is rotated 90° counterclockwise.

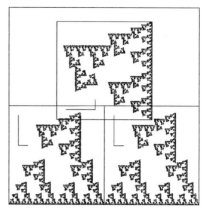

Figure 7. The resulting fractal produced by the MRCM of **Figure 6**, with the set of lenses overlaid.

The MRCM defined in **Figure 6** is identical to the Sierpinski Triangle MRCM except that the upper lens is rotated 90° counterclockwise.

Figure 7 shows the resulting fractal produced by the MRCM of **Figure 6**—quite different from the Sierpinski Triangle. **Figures 8–9** show the resulting fractal when the upper lens is rotated 180°. In **Figures 7** and **9**, the set of lenses has been overlaid on top of the attracting fractal.

Note again this remarkable property true of all our fractals: *The entire image is repeated exactly in each lens.* If you think about this property, it is paradoxical; but it is so. Each of these fractals is composed of smaller copies of itself. But then, each of the smaller copies is composed of smaller copies of themselves.

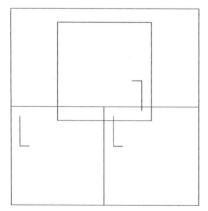

Figure 8. The resulting fractal when the upper lens is rotated 180°.

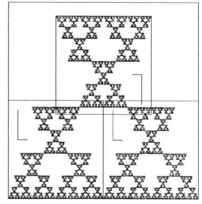

Figure 9. The resulting fractal when the upper lens is rotated 180°, with the set of lenses overlaid.

Thus, a fractal is composed of infinitely many copies of itself. You can think of the fractals produced by MRCMs as generalizations of the image you see when two mirrors face each other.

The term *self-similarity* of an image means that the image consists of scaled-down versions of itself. Later examples use lenses that are not squares. They may be parallelograms. In that case, the scaled down version of the entire image in that lens is the image distorted in the same way that the lens itself is the large square distorted.

The lenses need not be squares. A lens may have the shape of a rectangle, rhombus, or, more generally, any parallelogram. The entire attractor is composed of identical smaller copies of itself in each lens. If a lens is a rotated square, then the attractor will appear rotated in that lens. If a lens is a parallelogram, then the attractor will appear correspondingly deformed in that lens. In fact, the attractor is repeated not only in each lens but also in each iteration of each lens. **Figure 10** shows the fractal of **Figure 7** overlaid with the second and third iterations of the lenses. Note how the entire fractal appears in each of the smaller squares.

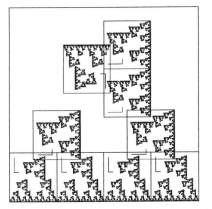

Figure 10a. The fractal of **Figure 7** overlaid with the second iteration of the lenses.

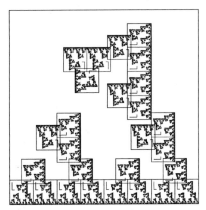

Figure 10b. The fractal of **Figure 7** overlaid with the third iteration of the lenses.

3. Specifying the Lenses

Each lens in an MRCM is a square that is changed in one or more of these ways:

- Size and shape: made smaller than the bounding square; deformed by stretching or contracting into a rectangle or parallelogram;

- Orientation: rotated and/or reflected;

- Location: shifted by translation so that it occupies a specific location in the large square.

7

Relative to the large square, how can the shape, orientation, and location of a parallelogram be described? **Figure 11** shows four parallelograms inside squares. Each has a specific size, shape, orientation, and location.

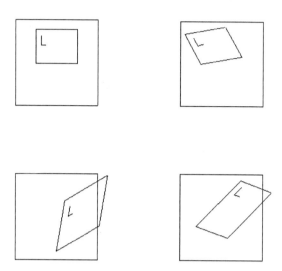

Figure 11. Four parallelograms inside squares, each with a specific size, shape, orientation, and location, as follows:

	e	f	r	s	ϕ	θ
a) upper left	25	50	50	40	0	0
b) upper right	25	50	50	40	10	30
c) lower left	50	10	60	60	30	−10
d) lower right	20	40	40	75	−20	−45

Six numbers (or parameters) e, f, r, s, ϕ, and θ are necessary to specify a lens. (These symbols are taken from Peitgen et al. [1992].) For convenience, we set the side of the large square to be 100 units. The lower-left corner of the large square is the coordinate origin $(0,0)$. In using the Java applet, a lens begins as the same shape and size as the large square. By changing the parameters, its shape, orientation, and location change.

r is the length of the horizontal side of the lens. After rotation, what was initially the horizontal side of the lens may no longer be horizontal; r is the length of what was the horizontal side before rotation. Similarly, s is the length of what was the vertical side of the lens before any rotation.

ϕ is the angle of inclination of the side labeled r relative to the horizontal axis. Similarly, the angle θ is the inclination of the side labeled s relative to the vertical axis. We measure these angles in degrees.

e and f are the x- and y-coordinates of the lower-left corner. The large square has its lower-left corner at the coordinate origin $(0, 0)$. However, if the lens is rotated or otherwise transformed, the original lower-left corner of the lens may not be what is now the lower-left corner; (e, f) is the location of the corner of the lens that began as the lower-left corner before rotations or other transformations.

The six parameters are best understood through examples.

4.   Examples

The following **Figures 12–21** show the MRCMs overlaid on top of their attracting fractals. The parameters of each lens are given. Study these; see if you can identify which parameters correspond with each lens. Also, check that most paradoxical property of self-similarity: that the entire fractal is repeated suitably distorted in each of the lenses so that the entire fractal consists of scaled-down versions of itself.

Figure 19 is due to Michael Barnsley (see references). This fern has become, if anything has, the emblem for iterated function systems. The stem is imaged by lens 4: The entire fern is collapsed into a vertical rectangle and placed at the location of the stem.

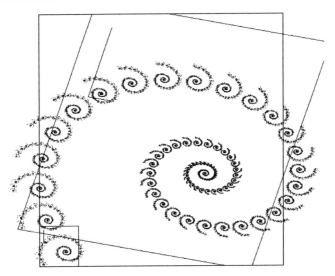

Figure 12.

lens	e	f	r	s	ϕ	θ
1	-9	15	97	96	-10	-20
2	0	0	16	16	0	0

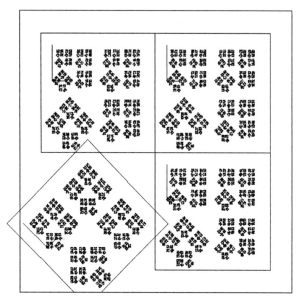

Figure 13.

lens	e	f	r	s	ϕ	θ
1	25	−5	42	42	45	45
2	50	8	42	42	0	0
3	50	50	42	42	0	0
4	8	50	42	42	0	0

Figures 19–21 show that natural forms can be modeled visually by fractals. The idea is intriguing. A natural form such as a tree has self-similarity: The maple tree from a distance has the same general shape, roughly, as the shape of each individual leaf. A water wave has curlicues at its extremities which bend over similar to the shape of the entire wave.

5.   Creating Fractals

There are two methods of creating fractals on the computer. The first is the MRCM, which I have used to present the idea of a fractal and self-similarity.

The second method is simpler to implement as a computer program and produces the fractal more quickly. It is called the *Fortune Wheel Reduction Copy Machine*. Here is how it works: Start with any point P_0. Choose one of the lenses at random. Transform the point through that lens. That is, think of the point as an image and run it through that lens. (In more detail, we think here of a lens as an operation that transforms points. The lens repositions a point to obtain a new point in the same way that lenses in the previous examples

10

Figure 14.

lens	e	f	r	s	ϕ	θ
1	50	0	70	70	45	45
2	0	0	25	25	0	0
3	75	0	25	25	0	0
4	0	75	25	25	0	0
5	75	75	25	25	0	0

change images.) Call this new point P_1. With P_1 repeat the same process: Choose a lens at random and run P_1 as an image through this lens. Continue. Eventually the plotted points will fill in the attractor. If the initial point P_0 is on the attractor, then all subsequent points will also be on the attractor. If P_0 is not on the attractor, then the first several points (perhaps a hundred or so) may not be close to the attractor. But eventually a point will land on the attractor—to the approximation of the pixel size—and from that point on, all subsequent points will reside on the attractor.

The fractal images displayed in this Module were made using the Fortune Wheel Reduction Copy Machine. A rather deep theorem is required to show that this method actually works—that is, that the resulting set of points approximates the actual attractor to closer and closer precision as more and more points are plotted. See the references for details of the proof.

In each stage of using the Fortune Wheel Reduction Copy Machine, a lens is chosen at random. More specifically, a weighting factor is required. Larger lenses must be chosen more often—in proportion to their area—than smaller

11

Figure 15.

lens	e	f	r	s	ϕ	θ
1	33	33	34	34	0	0
2	0	33	33	34	0	0
3	67	33	33	34	0	0
4	33	0	34	33	0	0
5	33	67	34	33	0	0

lenses.

The method by which a point image is run through a lens is this: A specific lens is defined by specific values of the six parameters e, f, r, s, ϕ, and θ as explained in **Section 2**. Given the point with coordinates (x, y), the new point (x_{new}, y_{new}) that results from applying the lens to (x, y) is

$$x_{new} = e + (r \cos \phi)\, x + (s \sin \theta)\, y, \qquad y_{new} = f - (r \sin \phi)\, x + (s \cos \theta)\, y.$$

The property of self-similarity means that the fractal is repeated inside itself—once in each lens—so that in fact it is composed of smaller copies of itself. All that you need to draw the fractal associated with a given MRCM is the lens system. It is as though the very process by which the fractal is created—the specific lens system defining it—is sufficient to completely describe the shape. Using the Fortune Wheel Reduction Copy Machine method, the specific points composing the final shape are random. Yet the fractal is completely specified regardless of the randomness used in the process of its construction.

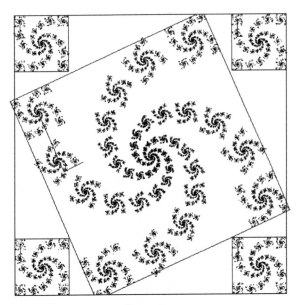

Figure 16.

lens	e	f	r	s	ϕ	θ
1	30	−2	78	78	24	24
2	0	0	20	20	0	0
3	100	0	20	20	90	90
4	100	100	20	20	180	180
5	0	100	20	20	−90	−90

In fact to draw a wave, for example, using an MRCM, you need not specify the position of each point (molecule) in the wave. That would take an enormous amount of information. You only need specify the lens system, and that requires rather little information.

This leads to a most useful application of MRCMs. For example, to transmit the shape of a mountain through a data channel, you only need specify the lens system. You don't need to specify the location of each rock, outcropping, and ravine. The lens system defining the mountain will create a shape that looks like the mountain you want to specify, although the specifics—the locations of individual rocks and such—will vary each time the fractal is constructed from the lens system. Of course, the lens system describing a particular mountain in the Appalachian range will differ from the lens system describing a Rocky mountain. Thus, MRCMs have applicability in the transmission of images over data lines.

13

Figure 17.

lens	e	f	r	s	ϕ	θ
1	20	20	60	60	0	0
2	0	41	18	18	0	0
3	11	17	18	18	30	30
4	33	2	18	18	60	60
5	59	0	18	18	90	90
6	83	11	18	18	120	120
7	98	32	18	18	150	150
8	100	58	18	18	180	180
9	90	82	18	18	−150	−150
10	68	98	18	18	−120	−120
11	41	100	18	18	−90	−90
12	17	90	18	18	−60	−60
13	2	68	18	18	−30	−30

Figure 18.

lens	e	f	r	s	ϕ	θ
1	0	50	75	75	−48	−42
2	0	0	36	36	0	0
3	100	100	36	36	180	180

6. Using the Program

The program used to create the figures in this Module is a Java applet on the Internet. You can access it by entering the following address into a Java-enabled browser such as Netscape:

```
http://www.warren-wilson.edu/~fsolomon/ZZFract.html
```

The first screen that you will encounter is the welcome screen. Click the mouse in the box to enter the Main Screen. Several boxes will be displayed:

- "View your lens system and/or change the lens setting"

- "Draw fractal using Multiple Reduction Copy Machine"

- "Draw fractal using Fortune Wheel Reduction Copy Machine"

- "Define an entirely new lens system"

- "Use a predefined example already stored in this applet"

15

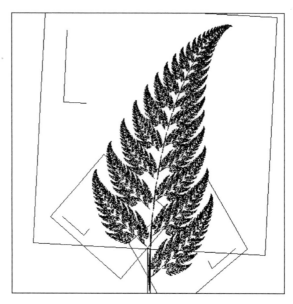

Figure 19. Barnsley's Fern.

lens	e	f	r	s	φ	θ
1	8	18	85	85	−3	−3
2	40	5	30	34	49	49
3	58	−8	30	37	120	−50
4	50	0	1	16	0	0

To get used to the program, you will probably want to begin by clicking the last box ("Use a predefined example ... "), which will take you to another screen: "List of Predefined Lens Systems." You can then select any of 12 lens systems already stored in the applet. Retrieve one by clicking the mouse on one of the boxes. You will find yourself back at the Main Screen with the data displayed for each of the lenses comprising the lens system that you selected. To see the lens system, click the "View your lens system and/or change the lens setting" box. You will then enter the Lens System screen. You can change any of the lenses or return to the Main Screen. Directions for changing, adding, and deleting lenses are provided.

From the Main Screen, you can:

- Draw your fractal in two ways, by selecting one of the two draw fractal boxes. Usually, you will want to use the Fortune Wheel Reduction Copy Machine. Select that box now. You will see a shape composed of 100 individual points. The Fortune Wheel Reduction Copy Machine (explained in **Section 4**) begins with the coordinate origin $(0, 0)$, iterates this point 100 times, and plots each point. Consequently, the image that you see contains 100 points. You can

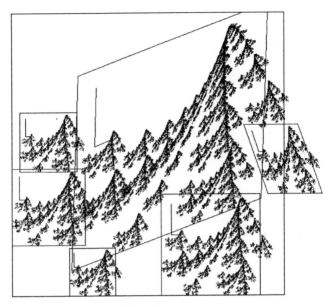

Figure 20.

lens	e	f	r	s	φ	θ
1	23	9	74	72	20	−2
2	55	0	36	36	0	0
3	3	44	21	21	0	0
4	21	0	17	17	0	0
5	0	18	27	27	0	0
6	94	36	20	26	0	20

click one of the other boxes to increase the number of points. Note that as the number of points increases, the fractal is filled in in more detail.

- Choose to draw the fractal using the Multiple Copy Reduction Machine, which follows the development that we have used in this Module. However, rather than beginning with the letter F, a large square is used at level 0.

- Define an entirely new lens system.

17

Figure 21.

lens	e	f	r	s	ϕ	θ
1	−2	26	82	82	−20	−20
2	0	8	30	30	−20	−20
3	100	8	30	30	200	20
4	0	67	24	24	0	0

7.   Exercises

1. For each part of this exercise, construct a square of side length 100 units on a piece of graph paper. (Select some useful scale so that the square occupies at least a quarter of the page.) Then construct the lens with each of these values for the parameters as explained in **Section 2**:

	e	f	r	s	ϕ	θ
a)	50	0	50	50	0	0
b)	80	60	20	30	0	0
c)	0	50	50	50	45	45
d)	50	0	40	60	0	0
e)	30	40	60	30	30	−30
f)	50	50	50	50	180	0
g)	40	60	50	50	−45	−45
h)	10	30	60	50	90	90
i)	20	50	30	10	90	−90

2. Determine whether you are correct by using the applet. From the Main Screen, click the leftmost box: "View your lens system and/or change the lens setting." Then modify the lens parameters to agree with each part of **Exercise 1**. Do your answers match the lenses constructed by the applet?

3. What values of the parameters e, f, r, s, ϕ, and θ are required to define each of the four lenses in **Figure 22**?

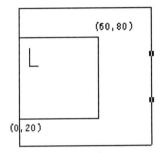

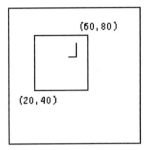

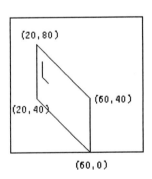

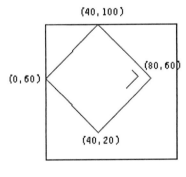

Figure 22. Figure for **Exercise 3**.

In **Exercises 4–8**, use the applet to construct the lens system with the given set of parameters. (From the Main Screen, click the "Define an entirely new lens system" box. Once the lens system is entered, return to the Main Screen and then click on the draw fractal using Fortune Wheel Reduction Copy Machine box to draw the fractal.) Before using the applet to draw the associated fractal, try to predict the visual shape of the fractal based on the appearance of the lens system and the property of self-similarity: that the entire fractal appears in each lens.

4.

lens	e	f	r	s	φ	θ
1	49	−18	97	97	45	45
2	0	0	18	18	0	0

5.

lens	e	f	r	s	φ	θ
1	0	0	50	50	0	0
2	50	0	50	50	0	0
3	0	50	50	50	0	0

6.

lens	e	f	r	s	φ	θ
1	42	50	50	50	−60	−60
2	56	42	50	50	60	60
3	56	58	50	50	180	180

7.

lens	e	f	r	s	φ	θ
1	18	31	37	37	−120	−60
2	68	18	37	37	−30	30
3	32	82	37	37	150	210
4	82	68	37	37	60	120
5	75	35	29	29	30	60
6	36	25	28	28	−60	−30
7	25	64	29	29	−150	−120
8	64	75	28	28	120	150

8.

lens	e	f	r	s	φ	θ
1	18	0	64	64	0	0
2	0	41	18	18	0	0
3	2	68	18	18	−30	−30
4	17	90	18	18	−60	−60
5	41	100	18	18	−90	−90
6	67	99	18	18	−120	−120
7	90	83	18	18	−150	−150
8	100	59	18	18	180	180

9. Enter the lens system of **Figure 12** into the applet by clicking the "Define an entirely new lens system" box from the Main Screen.

   a) Several times change the angles ϕ and θ of the large lens and observe the effect on the resulting fractal by returning to the Main Screen and then clicking the "Draw fractal using Fortune Wheel Reduction Copy Machine" box. To change the angles each time, click the "View your lens system and/or change the lens setting" box. What differences are there?

   b) Change the size of the small lens (do it several times) and observe the effects on the resulting fractal.

10. Enter the lens system of **Figure 16** into the applet. Construct the fractal using the Fortune Wheel Reduction Copy Machine to check that your fractal agrees with **Figure 16**. Now delete one of the corner lenses. How do you predict the resulting fractal will appear? Does your prediction agree with the actual fractal resulting from the applet? Now delete a second lens so that the two corner lenses are at opposite corners. Does your prediction agree with the fractal drawn by the applet? Now delete two of the original four corner lenses so that the remaining corner lenses are at adjacent corners; answer the same questions.

11. Enter the lens system of **Figure 16** into the applet. Several times change the angles of the large lens while maintaining $\phi = \theta$ (that is, maintain the large lens as a square). What differences are there in the resulting fractal?

12. Use the applet to define a lens system whose attracting fractal appears like a snowflake. Hint: Embody the hexagonal symmetry of the snowflake in the lens system. For example, define one lens at the center and six lenses symmetrically surrounding it. You will need to experiment to draw a snowflake shape rather than some other shape with hexagonal symmetry.

13. Define a lens system to draw some other natural form (other than a snowflake, that is). Examples: pine tree, maple tree, formation of flying geese, mountain, waterfall.

14. Repeat **Exercise 13** for another natural form.

15. Consider the parallelogram in **Figure 23**. Note that the angle ϕ is the angle of the side of length r with respect to the horizontal direction and angle θ is the angle of the side of length s with respect to the vertical direction. Prove that the area of the parallelogram is

$$A = rh = r\left(s\sin(\theta + 90° - \phi)\right) = rs\sin\left(90° - (\phi - \theta)\right) = rs\cos(\phi - \theta).$$

Consequently, the area of the lens with parameters e, f, r, s, ϕ, and θ is less than 1 if and only if $rs\cos(\phi - \theta) < 1$.

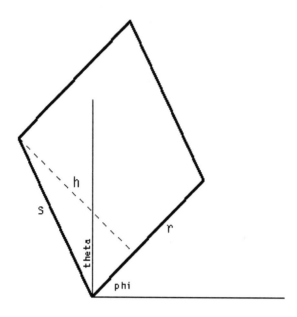

Figure 23. Figure for **Exercise 15**.

References

This Module has the sole purpose of showing how to use the Java applet to create fractals using MRCMs and understanding self-similarity. A full treatment of these ideas is developed in the first two books listed below. Fractals are also often associated with the Mandelbrot set, which is not the concern of this Module. Readers interested in the Mandelbrot set and related ideas might want to consult the third reference listed below. As you can imagine, there are many excellent Internet sites devoted to fractals and iterated function systems.

Peitgen, Heinz-Otto, Harmut Jergens, and Dietmar Saupe. 1992. *Fractals for the Classroom. Part One: Introduction to Fractals and Chaos.* New York: Springer-Verlag.

Barnsley, Michael. 1988. *Fractals Everywhere.* Boston, MA: Academic Press.

As a start, I highly recommend the first of the above books. It assumes only a background in basic algebra and geometry yet is thorough in its treatment. It has numerous examples, provides many insights and peripheral but related ideas. The second is at a much higher level of sophistication but covers in the first three chapters the theory developed in the entire first book. The author has been in the forefront of the development of the subject.

Devaney, Robert L. 1990. *Chaos, Fractals, and Dynamics: Computer Experiments in Mathematics.* Menlo Park, CA: Addison-Wesley.

Stewart, Ian. 1997. *The Magical Maze: Seeing the World Through Mathematical Eyes.* New York: Wiley.

Pp. 228–233 have a straightforward, concise explanation of the applications of iterated function systems to data compression.

Acknowledgments

Student Ginevra Clark during the summer 1997 devised many fractals, using a Pascal version of this applet. She created fractals mimicking waves, mountains, flocks of birds, galaxies, and many many others. Two of her creations appear as predefined examples in the applet. Ginevra and the author are most appreciative of a Student/Faculty Research Grant from the Appalachian College Association, which funded the summer project. Ginevra currently is a graduate student in chemistry at Brandeis University.

About the Author

Fred Solomon grew up in upstate New York, graduated from MIT, and earned a Ph.D. from Cornell University in 1972. After teaching at several colleges, he found his true home at Warren Wilson College in the North Carolina Blue Ridge Mountains, where he comprises half of the Mathematics Department. He spends much time hanging out with his Shetland sheepdog, who faithfully attends all Fred's classes but occasionally walks out early.

UMAP

Modules in
Undergraduate
Mathematics
and Its
Applications

Published in
cooperation with

The Society for
Industrial and
Applied Mathematics,

The Mathematical
Association of America,

The National Council
of Teachers of
Mathematics,

The American
Mathematical
Association of
Two-Year Colleges,

The Institute for
Operations Research
and the Management
Sciences, and

The American
Statistical Association.

Module 780

Applications of Sequences and Limits in Calculus

Yves Nievergelt

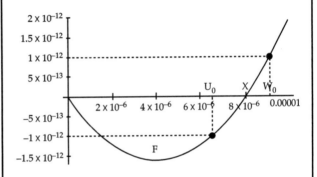

**Applications of Precalculus,
Calculus, and Analysis
to Chemistry and Finance**

COMAP, Inc., Suite 210, 57 Bedford Street, Lexington, MA 02420 (781) 862–7878

INTERMODULAR DESCRIPTION SHEET:	UMAP Unit 780

TITLE: Applications of Sequences and Limits in Calculus

AUTHOR: Yves Nievergelt
Dept. of Mathematics, MS 32
Eastern Washington University
526 5th Street
Cheney, WA 99004–2431
ynievergelt@mail.ewu.edu

MATHEMATICAL FIELD: Precalculus, calculus, and analysis

APPLICATION FIELD: Chemistry, finance

TARGET AUDIENCE: Students in precalculus, calculus, or analysis.

ABSTRACT: This Module introduces applications of the mathematical concepts of "sequence" and "limit" by way of the method of bisection. In precalculus, bisection provides an example of an algorithm, together with approximations of solutions, estimates of accuracy, sequences, and mathematical induction. In beginning calculus, bisection provides a motivation for studying in detail the convergence of sequences to limits. In advanced calculus and real analysis, bisection provides a motivation for precise definitions of the real numbers, for instance, as Dedekind cuts, or as equivalence classes of Cauchy sequences of rational numbers.

PREREQUISITES: A working knowledge of rational arithmetic, the concepts of mathematical functions, and mathematical induction.

RELATED UNITS: Unit 263:   *Horner's Scheme and Related Algorithms*, by Werner C. Rheinboldt. *Tools for Teaching 1980*, 497– 520. Boston, MA: Birkhäuser, 1981. Reprinted in *The UMAP Library*. CD-ROM. Lexington, MA: COMAP, 1996.

Unit 264:   *Algorithms for Finding Zeros of Functions*, by Werner C. Rheinboldt. Reprinted in *The UMAP Journal* 2 (1): 43–72. Reprinted in *Tools for Teaching 1981*, 471–500. Boston, MA: Birkhäuser, 1982. Reprinted in *The UMAP Library*. CD-ROM. Lexington, MA: COMAP, 1996.

Tools for Teaching 1999, 109–134. Reprinted from *The UMAP Journal* 20 (1) (1999) 139–164.
©Copyright 1999, 2000 by COMAP, Inc. All rights reserved.

COMAP, Inc., Suite 210, 57 Bedford Street, Lexington, MA 02420
(800) 77-COMAP = (800) 772-6627, or (781) 862-7878; http://www.comap.com

Applications of Sequences and Limits in Calculus

Yves Nievergelt
Dept. of Mathematics, MS 32
Eastern Washington University
526 5th Street
Cheney, WA 99004–2431
ynievergelt@mail.ewu.edu

Table of Contents

MODULES AND MONOGRAPHS IN UNDERGRADUATE
MATHEMATICS AND ITS APPLICATIONS (UMAP) PROJECT

The goal of UMAP is to develop, through a community of users and developers, a system of instructional modules in undergraduate mathematics and its applications, to be used to supplement existing courses and from which complete courses may eventually be built.

The Project was guided by a National Advisory Board of mathematicians, scientists, and educators. UMAP was funded by a grant from the National Science Foundation and now is supported by the Consortium for Mathematics and Its Applications (COMAP), Inc., a nonprofit corporation engaged in research and development in mathematics education.

Paul J. Campbell   Editor
Solomon Garfunkel   Executive Director, COMAP

1. Introduction

The mathematical concepts of "sequence" and "limit" are introduced in calculus texts of many genres as a technicality necessary for the presentation of such mathematical concepts as "continuity," "derivative," "integral," and "series" [Etgen 1999; Ostebee and Zorn 1997, 1998; Spivak 1980; Stewart 1995; Strang 1991]. Consequently, if the corresponding calculus course fails to provide extensive and intensive practice with limits, then students can earn high grades in calculus without having a clue about the nature and significance of the concept of a limit. As evidence of students' weak command of the concept of limit, an informal poll of students in second term calculus and linear algebra showed that over 93% of students do not solve the following exercise correctly:

1. Indicate whether the following statement is true or whether it is false:

   "For each function f defined in an open interval containing a number T,
   if the value of $f(t)$ gets closer and closer to a number L
   as the value of t gets closer and closer to the number T,
   then $\lim_{t \to T} f(t) = L$."

   □

Students' command of limits appears to strengthen with additional practice and demonstrations that the mathematical concept of limit is not a mere technicality. *Instead, the mathematical concept of limit is a major mathematical and philosophical accomplishment.* Indeed, since Archimedes' algorithm to compute π and the Babylonian and Chinese algorithms to extract square roots and cube roots [Neugebauer 1969, van der Waerden 1983], it took humanity over two millenia before Karl Weierstrass was able to state a precise definition (independent of any infinitesimal *deus ex machina*) of the mathematical concept of limit [Edwards 1979, 333]. The mathematical concept of limit seems to be the only concept from calculus that has such a major historical and philosophical status, and yet its detailed study is omitted in many courses. Mathematicians appear thus to form the only profession that has intentionally and successfully removed its major achievement from all of its introductory freshman courses, as evidenced by [Mac Lane 1997].

Of course, despite all the merits attributed to the use of history in pedagogy, dropping such famous names as Archimedes, Weierstrass, and the Chinese Han dynasty in such pompous sentences as the preceding ones will also fail to strengthen students' command of limits!

Therefore, these notes introduce the mathematical concepts of sequence and limit through a practical use that already contains the germs of the abstract idea: the method of bisection to solve equations with one unknown. These notes can be used before derivatives and integrals in the first term of calculus; or before series in subsequent terms of calculus, depending on the syllabus; or as a fast motivation or review for students in beginning real analysis.

1

2. Reasons for Solving Equations

The need to solve equations arises because some numbers are defined implicitly ("abstractly") as the solutions of equations. For some equations, there exists no formula to calculate the solutions, or the existing formula merely gives a name to the solution but does not provide a method to calculate it. In either case, the computation of the solutions can involve iterative methods, as is the case for the bisection method presented here. In general, iterative methods produce infinite sequences of numbers, which can approximate solutions of equations to any specified accuracy. Moreover, well-designed iterative methods can prove more effective than exact but ill-suited formulae, especially with the finite arithmetic of digital computers.

Example 1. The equation $3x = 2$ defines the number $x = 2/3$. Yet the formula $2/3$ merely gives a name to the solution x of $3x = 2$, without providing its decimal expansion. Nevertheless, the iterative method of long division produces an infinite sequence of numbers 0.6, 0.66, 0.666, $\ldots$, which yields the decimal expansion $2/3 = 0.666\,666\ldots$. (However, the formula $2/3$ does contain the expansion of x in base 3: $x = 2/3 = 2*3^{-1} = 0.2_{\text{three}}$.) $\square$

Example 2. Consider the problem of computing the length d of the diagonal of a square with all sides of length 1 (see **Figure 1**).

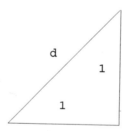

Figure 1. $d^2 = 1^2 + 1^2$.

The only information about d provided by geometry comes from the Pythagorean theorem, which states that $d^2 = 1^2 + 1^2$ but does not produce digits for the length d of the diagonal. The bisection method, among many other iterative methods, will produce the decimal expansion of d. $\square$

Exercises

2. Write down an equation involving only integer arithmetic (polynomial) with $\sqrt{3}$ as one of its solutions.

3. Write down an equation involving only integer arithmetic (polynomial) with $\sqrt[3]{2}$ as one of its solutions.

3. Solving Equations through Bisection

The method of bisection is perhaps the simplest, most reliable, and most general algorithm to solve equations. Though the method of bisection does not instantly produce the solution exactly, it refines the accuracy at each step and can achieve any specified accuracy. Moreover, the method of bisection is a simple prototype for *iterative algorithms*, which form the basis of most of scientific computing.

To solve an equation of the form $f(x) = 0$, the method of bisection requires only two initial estimates $u_0 \le w_0$ with

$$f(u_0) \le 0 \le f(w_0),$$

as in **Figure 2**, from which successive bisections will produce a solution in the closed (including both endpoints) interval $[u_0, w_0]$.

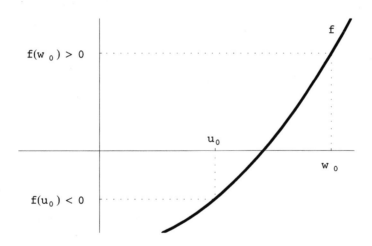

Figure 2. Bisection yields a solution in $[u_0, w_0]$ if $f(u_0) \le 0 \le f(w_0)$.

To narrow the interval, the method tests the value of f at the midpoint

$$v_0 := \frac{u_0 + w_0}{2},$$

where two cases may arise. (The symbol := defines the yet undefined left-hand side in terms of the already defined right-hand side.)

- (+) If $f(v_0) \ge 0$, then

$$f(u_0) \le 0 \le f(v_0), f(w_0),$$

3

and the method narrows the interval to the left-hand half $[u_1, w_1] := [u_0, v_0]$.

- $(-)$ If $f(v_0) < 0$, then

$$f(u_0), f(v_0) \leq 0 \leq f(w_0),$$

and the method narrows the interval to the right-hand half $[u_1, w_1] := [v_0, w_0]$.

In either case, $f(u_1) \leq 0 \leq f(w_1)$ with a narrower interval $[u_1, w_1]$.

To solve a problem through the method of bisection as just described, a restatement of the problem into an equation of the form $f(x) = 0$ may first be necessary.

Example 3. To compute the length d of the diagonal of a square with sides of length 1, first rewrite d as the positive solution of the equation $x^2 = 2$, and then rewrite the equation as $x^2 - 2 = 0$ so that $f(x) = x^2 - 2$.

Then select two initial estimates — perhaps after some trial and error — for instance, as in **Figure 3**:

$$u_0 := 1, \quad f(u_0) = f(1) = 1^2 - 2 = -1 < 0,$$
$$w_0 := 2, \quad f(w_0) = f(2) = 2^2 - 2 = 2 > 0.$$

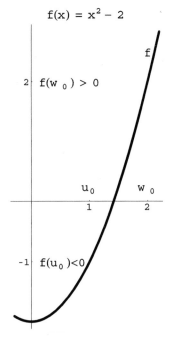

$$f(x) = x^2 - 2$$

Figure 3. Initial estimates $[u_0, w_0] = [1, 2]$ to solve $x^2 - 2 = 0$.

Step 1. The method of bisection then tests f at the midpoint v_0:

$$v_0 := \frac{u_0 + w_0}{2} = \frac{1 + 2}{2} = 3/2,$$
$$f(v_0) = f(3/2) = (3/2)^2 - 2 = 9/4 - 2 = 1/4 > 0.$$

Consequently,

$$f(u_0) < 0 < f(v_0), f(w_0),$$

so the shorter interval $[u_0, v_0] = [1, 3/2]$ may replace the longer initial interval $[u_0, w_0] = [1, 2]$; set

$$u_1 := u_0 = 1, \quad w_1 := v_0 = 3/2.$$

Thus, the first step yields the result $[u_1, w_1] = [1, 3/2]$.

Step 2. The method of bisection then proceeds by testing f at the new midpoint v_1:

$$v_1 := \frac{u_1 + w_1}{2} = \frac{1 + 3/2}{2} = 5/4,$$
$$f(v_1) = f(5/4) = (5/4)^2 - 2 = 25/16 - 2 = -7/16 < 0.$$

Consequently,

$$f(u_1), f(v_1) < 0 < f(w_1),$$

so the shorter interval $[v_1, w_1] = [5/4, 3/2]$ may replace the longer initial interval $[u_1, w_1] = [1, 3/2]$; set

$$u_2 := v_1 = 5/4, \quad w_2 := w_1 = 3/2.$$

Thus, the second step yields the result $[u_2, w_2] = [5/4, 3/2]$.

Similarly, the next few steps produce the results of **Table 1**.

Table 1.

Steps in the calculation, by means of bisection, of the length of the diagonal of a unit square.

n	u_n	v_n	w_n
0	1	1.5	2
1	1	1.25	1.5
2	1.25	1.375	1.5
3	1.375	1.4375	1.5
4	1.375	1.40625	1.4375
5	1.40625	1.4140625	1.4375
6	1.4140625	1.41796875	1.421875
7	1.4140625	1.416015625	1.41796875
8	1.4140625	1.4150390625	1.416015625
9	1.4140625	1.41455078125	1.4150390625
10	1.4140625	1.414306640625	1.41455078125
⋮	⋮	⋮	⋮

The last line displayed shows that $1.4140 < d < 1.4146$; thus, the bisection method has produced the first four decimal digits: $d = 1.414\ldots$.

□

In practice, the question arises how to determine in advance how many steps, how many arithmetic operations, or how much computing time will deliver a specified accuracy. The following considerations explain how to determine in advance how many steps of the bisection method will deliver a specified accuracy. In general, each single step of the method of bisection starts from two initial estimates $u_n \leq w_n$ with

$$f(u_n) \leq 0 \leq f(w_n).$$

To narrow the interval, the method tests the value of f at the midpoint

$$v_n := \frac{u_n + w_n}{2},$$

where two cases may arise.

- (+) If $f(v_n) \geq 0$, then

$$f(u_n) \leq 0 \leq f(v_n), f(w_n),$$

  and the interval narrows to the left-hand half $[u_{n+1}, w_{n+1}] := [u_n, v_n]$.

- (−) If $f(v_n) < 0$, then

$$f(u_n), f(v_n) \leq 0 \leq f(w_n),$$

  and the interval narrows to the right-hand half $[u_{n+1}, w_{n+1}] := [v_n, w_n]$.

In either case, $f(u_{n+1}) \leq 0 \leq f(w_{n+1})$ with a narrower interval $[u_{n+1}, w_{n+1}]$.

The method of bisection thus produces sequences of numbers with features described in the following proposition, the proof of which illustrates the concept of mathematical induction at the level of precalculus, for instance, at the level of Hungerford and Mercer [1985, 517–526].

Proposition 1. *For each function* f *defined on a closed interval* $[u_0, w_0]$ *where* $u_0 \leq w_0$ *and*

$$f(u_0) \leq 0 \leq f(w_0),$$

the method of bisection produces sequences of numbers (u_n) *and* (w_n) *such that*

$$u_0 \leq \cdots \leq u_n \leq u_{n+1} \leq \cdots \leq w_{n+1} \leq w_n \leq \cdots \leq w_0,$$

$$f(u_n) \leq 0 \leq f(w_n),$$

$$|w_n - u_n| = \frac{|w_0 - u_0|}{2^n}.$$

6

Proof: Because bisection proceeds inductively, the proof proceeds by induction. By hypothesis,

$$u_0 \leq w_0,$$
$$f(u_0) \leq 0 \leq f(w_0),$$
$$w_0 - u_0 = (w_0 - u_0)/2^0.$$

Hence, assume that for some nonnegative integer n,

$$u_n \leq w_n,$$
$$f(u_n) \leq 0 \leq f(w_n),$$
$$w_n - u_n = (w_0 - u_0)/2^n.$$

Step n. Then $u_n \leq v_n \leq w_n$, because

$$u_n = \frac{u_n + u_n}{2} \leq \frac{u_n + w_n}{2} = v_n = \frac{u_n + w_n}{2} \leq \frac{w_n + w_n}{2} = w_n.$$

Two cases may arise:

- If $f(v_n) \geq 0$, then $w_{n+1} := v_n$ and the preceding inequalities give

$$u_{n+1} = u_n \leq v_n = w_{n+1},$$

whence

$$u_0 \leq \cdots \leq u_n \leq u_{n+1} \leq v_n \leq w_{n+1} \leq w_n \leq \cdots \leq w_0,$$

$$f(u_{n+1}) = f(u_n) \leq 0 \leq f(v_n) = f(w_{n+1}),$$

$$w_{n+1} - u_{n+1} = v_n - u_n$$
$$= \frac{u_n + w_n}{2} - u_n = \frac{w_n - u_n}{2} = \frac{(w_0 - u_0)/2^n}{2} = \frac{w_0 - u_0}{2^{n+1}}.$$

- Whereas if $f(v_n) < 0$, then $u_{n+1} := v_n$ and

$$u_{n+1} = v_n \leq w_n = w_{n+1},$$

whence

$$u_0 \leq \cdots \leq u_n \leq u_{n+1} \leq v_n \leq w_{n+1} \leq w_n \leq \cdots \leq w_0,$$

$$f(u_{n+1}) = f(v_n) \leq 0 \leq f(w_n) = f(w_{n+1}),$$

$$w_{n+1} - u_{n+1} = w_n - v_n$$
$$= w_n - \frac{u_n + w_n}{2} = \frac{w_n - u_n}{2} = \frac{(w_0 - u_0)/2^n}{2} = \frac{w_0 - u_0}{2^{n+1}}.$$

In either case,

$$u_0 \leq \cdots \leq u_n \leq u_{n+1} \leq v_n \leq w_{n+1} \leq w_n \leq \cdots \leq w_0,$$

$$f(u_{n+1}) \leq 0 \leq f(w_{n+1}),$$

$$w_{n+1} - u_{n+1} = \frac{w_0 - u_0}{2^{n+1}}. \qquad \square$$

Example 4. Consider the problem of determining in advance the number n of steps sufficient (though perhaps not necessary) to compute to twelve significant digits the solution d of the equation $x^2 - 2 = 0$ from $u_0 := 1$ and $w_0 := 2$. Before starting the bisection algorithm, we know that the formula

$$w_n - u_n = \frac{w_0 - u_0}{2^n} = \frac{2 - 1}{2^n} = \frac{1}{2^n}$$

expresses the accuracy that will be achieved at the n th step. Thus, the problem amounts to solving for n the equation

$$(1/2) * 10^{-12} * d \geq w_n - u_n = 2^{-n}.$$

Because d remains unknown, but $1 < d < 2$, it suffices to solve

$$(1/2) * 10^{-12} * d > (1/2) * 10^{-12} * 1 \geq w_n - u_n = 2^{-n}.$$

Hence

$$2^{n-1} \geq 10^{12}.$$

Logarithms to base ten would give the solution in the form $n - 1 \geq 12 \log(10)/ \log(2) = 12/ \log(2) = 39.863 \ldots$, so that $n \geq 41$ suffices. However, computing a logarithm can demand more computational resources than computing d. One way to find such a value of n without logarithms is to compute powers of 2 until one of them exceeds 10^{12}, for instance, $2, 2^2, 2^4, 2^8 = 256, 2^{16}, 2^{32} > 4*10^9, 2^{32}*2^8 > 4*10^9*256 > 10^{12}$, whence $n - 1 = 32 + 8 = 40$ and $n = 41$ suffices. A faster but less conservative method is to use the inequality $2^4 > 10$ to simplify the problem $2^{n-1} \geq 10^{12}$ into the form

$$2^{n-1} \geq (2^4)^{(n-1)/4} > 10^{(n-1)/4} \geq 10^{12}$$

whence $(n - 1)/4 \geq 12$ and finally $n \geq 49$. This faster method involves neither logarithms nor powers of two, but it leads to $49 - 41 = 8$ additional unnecessary steps. $\qquad \square$

Exercises

The following four exercises provide practice using the method of bisection.

4. Apply the method of bisection to compute $\sqrt{3}$.

5. Apply the method of bisection to compute $\sqrt[3]{2}$.

6. Determine a number n of steps sufficient to compute $\sqrt{3}$ by the method of bisection to twelve significant digits.

7. Determine a number n of steps sufficient to compute $\sqrt[3]{2}$ by the method of bisection to twelve significant digits.

The following two exercises show results from the method of bisection applied to situations that do not satisfy the conditions stated for the method of bisection.

8. Apply the method of bisection to

$$f(x) = \frac{x^4 + x^2 - 6}{x^4 - 5x^2 + 6}$$

with $u_0 = 1$ and $w_0 = 2$.

9. Apply the method of bisection to

$$g(x) = \frac{5x^4 - 3x^3 - 3x^2 - 3x - 8}{x^4 - 5x^2 + 6}$$

with $u_0 = 1$ and $w_0 = 2$.

The following four exercises demonstrate the solution of equations in applications.

10. Apply the method of bisection to compute the positive solution v_* of the equation

$$10{,}193.75v^4 + 193.75v^3 + 193.75v^2 + 193.75v - 9{,}987.50 = 0.$$

This equation gives the internal rate of return $r_* = 2[(1/v_*) - 1]$ of the two-year note issue by the U.S. Treasury on 31 August 1993, in terms of the purchase price $9,987.50 on the day of issue, the semi-annual amount of interest, $193.75, and the value redeemed at maturity, $10,000.00 [*Wall Street Journal* 1993].

11. Apply the method of bisection to compute the positive solution X of the equation

$$X^3 + 0.1000269X^2 - 8.070\,000\,100\,7 \times 10^{-7}X - 2.70883 \times 10^{-19} = 0.$$

In chemistry, X represents the equilibrium concentration of hydrogen ions $[H^+]$ in a water solution of monoprotic acid, $H(CH_3CO_2)$, found in vinegar, and sodium acetate, $Na(CH_3CO_2)$, the "salt" obtained by reducing the acid. The quantity $pH = -\log_{10}([H^+]) = -\log_{10} X$ represents the acidity of the solution [MacLeod 1984].

12. Determine a number n of steps sufficient to compute the solution v_* of

$$10{,}193.75v^4 + 193.75v^3 + 193.75v^2 + 193.75v - 9{,}987.50 = 0$$

by the method of bisection to twelve significant digits. Your answer may depend on your initial values u_0 and w_0.

13. Determine a number n of steps sufficient to compute the solution X of

$$X^3 + 0.1000269X^2 - 8.070\,000\,100\,7 \times 10^{-7}X - 2.70883 \times 10^{-19} = 0$$

by the method of bisection to twelve significant digits. Your answer may depend on your initial values U_0 and W_0.

The following two exercises demonstrate issues arising in the design of algorithms.

14. Investigate the consequences of modifying the test so that either $f(v_n) > 0$ or $f(v_n) \le 0$, instead of $f(v_n) \ge 0$ or $f(v_n) < 0$.

15. Investigate the consequences of requiring that $f(u_0) < 0 < f(w_0)$ instead of $f(u_0) \le 0 \le f(w_0)$.

4. Further Issues

The preceding considerations raise a first and fundamental issue, about the precise nature of the "number" d. First, d is not a rational number.

Proposition 2. *There is no rational number d such that $d^2 = 2$.*

Proof: The main resource for this proof is the Fundamental Theorem of Arithmetic and the associated unique prime factorization, borrowed from algebra [Birkhoff and Mac Lane 1991, 23]. If $d^2 = 2$ and $d = p/q$ is rational, with relatively prime integers p and q, then $2 = (p/q)^2$, whence $2q^2 = p^2$. The unique prime factorizations of $p = p_1^{r_1} \cdots p_k^{r_k}$ and $q = q_1^{s_1} \cdots q_\ell^{s_\ell}$ then would give $2q_1^{2s_1} \cdots q_\ell^{2s_\ell} = 2q^2 = p^2 = p_1^{2r_1} \cdots p_k^{2r_k}$, where the left-hand side has an odd power of 2 while the right-hand side has an even power of 2, so that they cannot equal each other.   □

Therefore, discussing the "number" d requires numbers different from the rational numbers. What are such numbers? Essentially, the answer lies right in the numerical experiments from the preceding examples. All a user ever sees from d are sequences of rational numbers such as (u_n) and (w_n). Thus, d corresponds to all the sequences produced by the bisection method (or any other algorithm) in computing the nonnegative solution of $x^2 - 2 = 0$; the collection of all such sequences could be denoted by $\sqrt{2}$. This description of $\sqrt{2}$ in terms of a collection of sequences is similar to the description of a rational number $2/3$ in terms of such equivalent fractions as $4/6$ or $-14/-21$. Such ways to define real numbers are called *equivalent Cauchy sequences* and are outlined in Spivak [1980, 565].

Alternatively, because all that a user ever sees from d are either rationals p/q such that $(p/q)^2 < 2$ or rationals p/q such that $2 < (p/q)^2$, the "number" $\sqrt{2}$ can be defined as the set of all positive rationals p/q such that $2 < (p/q)^2$. All that a digital computer ever produces for $\sqrt{2}$ is either a rational number from that set, with $2 < (p/q)^2$, or a positive rational number from its complement, with $(p/q)^2 < 2$. Such ways to define real numbers are called *Dedekind cuts* and are described in detail in the original source [Dedekind 1963] and in the *calculus* text [Spivak 1980, 555].

Thus, whether the results from the method of bisection "converge" toward a solution of the equation $f(x) = 0$ depends on both the function f and the type of "numbers" allowed by the problem under consideration. For example, with $f(x) = x^2 - 2$ and real numbers in $[u_0, w_0] = [1, 2]$, the sequences (u_n) and (w_n) "converge" to the solution $\sqrt{2}$. In contrast, working with only rational numbers produces the same sequences (u_n) and (w_n) —but they fail to converge, because $\sqrt{2}$ is not a rational number.

The mathematical concept of "convergence" of a sequence does *not* mean that the terms of the sequence get closer and closer to the limit, but, rather, that they eventually get and remain closer than any specified tolerance.

Definition 1. A sequence (v_n) **converges** to a **limit** v_* if and only if

for every $\varepsilon > 0$ (the "tolerance")
there exists an integer m such that
$|v_n - v_*| < \varepsilon$ for every integer $n > m$.   □

Example 5. In computing the positive solution d of the equation $x^2 - 2 = 0$ from the initial estimates $u_0 := 1$ and $w_0 := 2$, the method of bisection produces three sequences, (u_n), (v_n), and (w_n), which all three converge to the limit $\sqrt{2}$ in the real numbers. Indeed,

for every $\varepsilon > 0$,
there exists an integer m such that $\varepsilon > 2^{-m} > 0$,
whence for every $n > m$, we have
$|u_n - v_*|, \ |v_n - v_*|, \ |w_n - v_*| \leq w_n - u_n < 2^{-n} < \varepsilon$.

The justification of the statement that "for every $\varepsilon > 0$ there exists an integer m such that $\varepsilon > 2^{-m} > 0$" depends on the definition of the real numbers adopted in the course.   □

Another issue pertains to multiple zeroes. If a function f has more than one zero inside an interval $[u_0, w_0]$, then the method of bisection converges to *some* zero of f, which zero depends on f, as seen in Benjamin [1987]. Moreover, the rate at which the method of bisection converges does not depend on the function f but depends only on the binary expansion of the selected zero [Nievergelt 1995].

Yet another issue relates to simultaneous systems of equations. Systems of equations with more than one real unknown require more sophisticated methods specialized to individual systems. For one complex polynomial equation with one complex unknown, which constitutes a special case of a system of two equations with two real unknowns, an algorithm of Herman Weyl's works similarly to the method of bisection, as explained by Victor Pan [1997].

Still another issue relates to the type of equation that defines a number. Some numbers cannot be defined by polynomial equations and require equations involving nonalgebraic functions. One such number is the ratio π of the circumference to the diameter of a circle [Lang 1965, 493–494].

A different issue relates to the accuracy obtainable in practice. In practice, the accuracy of the results depends on the accuracy of the computations of the values $f(v_n)$, in particular, whether the computer carries enough digits to determine which of $f(v_n) \geq 0$ or $f(v_n) < 0$ holds. Eventually, a computer's rounding can yield the wrong sign.

Exercises

16. Define the positive solution of $x^3 = 2$ in terms of Dedekind cuts.

17. This exercise pertains to the issue of computability.

   a) Prove that computations carried out to exactly ten decimal digits (rounding all intermediate arithmetic results to ten significant decimal digits) erroneously indicate that the following quadratic equation has no real solutions:

   $$0.3\,124\,999\,999 * Z^2 - 0.707\,106\,781\,1 * Z + 0.4 = 0.$$

   b) Prove that the following quadratic equation has two distinct real solutions:

   $$0.3\,124\,999\,999 * Z^2 - 0.707\,106\,781\,1 * Z + 0.4 = 0.$$

5.   Solving Equations through *Regula Falsi*

The purpose of this section is to allow students to study on their own an algorithm not presented in class, as may happen after they leave school. To this end, this section outlines the algorithm and leaves the investigations to the exercises.

The method of *false position*, also called by its Latin name *regula falsi*, proceeds as does the method of bisection, except for a different choice of the intermediate points v_n, which need no longer lie at the midpoint of $[u_n, w_n]$. To solve an equation of the form

$$f(x) = 0,$$

regula falsi requires only two initial estimates $u_0 < w_0$ with

$$f(u_0) < 0 < f(w_0).$$

For a more convenient notation, define

$$p_0 := f(u_0),$$
$$q_0 := f(w_0),$$

so that the graph of f passes through the two points

$$(u_0, p_0), \qquad (w_0, q_0),$$

with $p_0 < 0 < q_0$. Therefore, the straight line through (u_0, p_0) and (w_0, q_0) crosses the horizontal axis at some point $(v_0, 0)$ with $u_0 < v_0 < w_0$. *Regula falsi* selects this point v_0.

Then, exactly as for the method of bisection, at least two cases may arise.

- (+) If $f(v_0) > 0$, then

$$f(u_0) \leq 0 \leq f(v_0), f(w_0),$$

  and the interval narrows to the left-hand half $[u_1, w_1] := [u_0, v_0]$.

- (−) If $f(v_0) < 0$, then

$$f(u_0), f(v_0) \leq 0 \leq f(w_0),$$

  and the interval narrows to the right-hand half $[u_1, w_1] := [v_0, w_0]$.

In either case, $f(u_1) \leq 0 \leq f(w_1)$ with a narrower interval $[u_1, w_1]$. *Regula falsi* then iterates the same steps, finding a point v_1 between u_1 and w_1, and so on.

Exercises

18. Design a way to specify what the algorithm does for the yet unexamined cases $f(u_0) = 0$, $f(v_0) = 0$, and $f(w_0) = 0$.

19. Write an equation $y = mv + b$ for the straight line through two points (u_0, p_0) and (w_0, q_0), and solve for the value v_0 where the line crosses the horizontal axis: this gives a formula for v_0 in terms of u_0, p_0, w_0, q_0.

20. Imagine that you are a newly hired junior member of a scientific computing team and that you have just received the following assignment, with results to be reported to your boss: Locate a source in the library or elsewhere that proves that *regula falsi* converges to a solution of $f(x) = 0$. You may look in the index of books in the category called "numerical analysis" under Library of Congress Classification Numbers ("call numbers") QA297 ..., where a similar (but not quite as good) method called "secant method" may also appear. You need not reproduce the content of the proof here. Just name the source and report the page numbers, the authors' full names, the full book title, publisher, year, and the location in the library (such as Dewey Decimal Classification Number or Library of Congress Classification Number), so that on the sole basis of your report your boss can find this proof very quickly.

21. Apply a few steps of *regula falsi* to compute $\sqrt{3}$.

22. Prove that each of the sequences (u_n) and (w_n) produced by *regula falsi* converges to some limit in the real numbers.

23. Apply the method of *regula falsi* to compute the nonnegative solution X of

$$X^3 + 0.1000269X^2 - 8.070\,000\,100\,7 \times 10^{-7}X - 2.70883 \times 10^{-19} = 0.$$

(See also **Exercise 11**.)

24. If you have access to a calculator or a computer with a built-in program or software to solve equations, try it to compute the nonnegative solution X of

$$X^3 + 0.1000269X^2 - 8.070\,000\,100\,7 \times 10^{-7}X - 2.70883 \times 10^{-19} = 0.$$

(See also **Exercise 11**.)

6.   Conclusions

The mathematical concepts of sequence and limit are not only technicalities necessary for other concepts; they are also concepts that arise in scientific computing, and they form the basis for all the other concepts from calculus, including derivatives, integrals, inverse functions, and infinite series. Moreover, the mathematical concept of limit appears to be the major philosophical achievement of mathematics presented, distorted, or withheld, in calculus.

7. Solutions to the Odd-Numbered Exercises

1. The following statement is *false*: "For each function f defined in an open interval containing a number T, if the value of $f(t)$ gets closer and closer to a number L as the value of t gets closer and closer to the number T, then $\lim_{t \to T} f(t) = L$." Indeed, consider the function f defined by $f(t) = 1 - t^2$, for instance, with $T = 0$ and $L = 3$. Then $f(t)$ gets closer and closer to 3 as t gets closer and closer to 0, but the limit is 1, *not* 3.

   The concept of limit does not merely require that $f(t)$ get closer and closer to L; it requires that for each positive specification $s > 0$ the value of $f(t)$ gets within s of the limit L for every value of t within some positive tolerance $r > 0$ of the point T:

   for each $s > 0$,
   there exists some $r > 0$ (which may depend on f, T, and s) such that
   $|f(t) - L| < s$
   for every t with $0 < |t - T| < r$.

   For the particular example with $f(t) = 1 - t^2$ and $T = 0$, the value of $f(t)$ never gets within $s = 1$ of 3; indeed, while $f(t)$ gets closer and closer to 3, it always remains at least 2 away.

3. $x^3 = 2$.

5. To solve the equation $x^3 - 2 = 0$ with $g(x) := x^3 - 2$ through the method of bisection, for instance, starting with $u_0 := 1$ and $w_0 := 2$, the first few steps produce the results of **Table 2**.

Table 2.

Steps in the calculation, by means of bisection, of the real cube root of 2.

n	u_n	v_n	w_n
0	1	1.5	2
1	1	1.25	1.5
2	1.25	1.375	1.5
3	1.25	1.3125	1.375
4	1.25	1.28125	1.3125
5	1.25	1.265625	1.28125
6	1.25	1.2578125	1.265625
7	1.2578125	1.26171875	1.265625
8	1.2578125	1.259765625	1.26171875
9	1.259765625	1.2607421875	1.26171875
10	1.259765625	1.26025390625	1.2607421875
⋮	⋮	⋮	⋮

The last line of the table shows that $1.259765625 < \sqrt[3]{2} < 1.2607421875$.

15

7. Proceed as in **Example 4**, with the same result $n \geq 41$.

9. From $u_0 = 1$ and $w_0 = 2$, the method of bisection converges to $\sqrt{2}$, which is *not* a solution of $g(x) = 0$ but instead marks a vertical asymptote of g:

$$g(x) = \frac{5x^4 - 3x^3 - 3x^2 - 3x - 8}{x^4 - 5x^2 + 6} = \frac{5(x - 1.6)(x + 1)(x^2 + 1)}{(x^2 - 2)(x^2 - 3)}.$$

11.

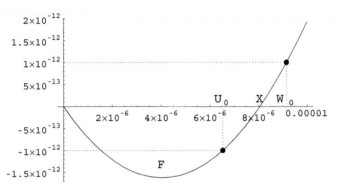

Figure 4. $F(X) = X^3 + 0.1000269X^2 - 8.070\,000\,100\,7 \times 10^{-7}X - 2.70883 \times 10^{-19}$.

See **Figure 4**. Substitutions show that $F(0) = -2.70883 * 10^{-19} < 0 < 1 - 2.70883 * 10^{-19} < F(1)$. Consequently, $U_0 := 0$ and $W_0 := 1$ satisfy the requirements, but from such initial values the bisection requires dozens of steps to produce at least one exact significant digit, showing that $8.0 * 10^{-6} < X < 8.1 * 10^{-6}$. Additional steps give the approximation $X \approx 8.067 * 10^{-6}$. See also **Figure 4**.

13. Because $8.0 * 10^{-6} < X < 8.1 * 10^{-6}$, it follows that for an accuracy of twelve significant digits it suffices that

$$(1/2) * 10^{-12} * X > (1/2) * 10^{-12} * 8.0 * 10^{-6} = 4 * 10^{-18} \geq (W_0 - U_0)/2^n.$$

With $U_0 := 0$ and $W_0 := 1$, these inequalities lead to $2^{n+2} = 2^n * 4 \geq 10^{18}$, whence $n \geq 58$ suffices.

15. With $f(u_0) < 0 < f(w_0)$, it can happen that $f(v_n) = 0$ at some later stage $n > 0$. This does not cause any difficulty, but it requires a special case in the bisection algorithm, for instance, stopping the algorithm, or setting $u_k := w_k := v_n$ for every $k > n$.

17. a) For the quadratic equation

$$0.3\,124\,999\,999 * Z^2 - 0.707\,106\,781\,1 * Z + 0.4 = 0,$$

hand calculations rounding every result to ten significant decimal digits or a ten-digit decimal calculator (HP-15C) computes the discriminant as follows, where $\texttt{float}(r)$ denotes the result of computing r to exactly ten significant decimal digits.

$$\begin{aligned}
\texttt{float}(b^2) &= \texttt{float}([-0.707\,106\,781\,1]^2) \\
&= 0.499\,999\,999\,9; \\
\texttt{float}(a * c) &= \texttt{float}(0.3\,124\,999\,999 * 0.4) \\
&= 0.125\,000\,000\,0; \\
\texttt{float}(4 * a * c) &= \texttt{float}(4 * 0.3\,124\,999\,999 * 0.4) \\
&= 0.500\,000\,000\,0; \\
\texttt{float}(b^2 - 4 * a * c) &= \texttt{float}([-0.707\,106\,781\,1]^2 - 4 * 0.3\,124\,999\,999 * 0.4) \\
&= 0.499\,999\,999\,9 - 0.500\,000\,000\,0 \\
&= -0.000\,000\,000\,1 \\
&< 0.
\end{aligned}$$

This negative value of the computed discriminant would indicate that the proposed quadratic equation has no real solution.

b) Yet exact calculations reveal that

$$\begin{aligned}
b^2 &= [-0.707\,106\,781\,1]^2 \\
&> 0.499\,999\,999\,877; \\
a * c &= 0.3\,124\,999\,999 * 0.4 \\
&= 0.124\,999\,999\,96; \\
4 * a * c &= 4 * 0.3\,124\,999\,999 * 0.4 \\
&= 0.499\,999\,999\,84; \\
b^2 - 4 * a * c &= [-0.707\,106\,781\,1]^2 - 4 * 0.3\,124\,999\,999 * 0.4 \\
&> 0.499\,999\,999\,877 - 0.499\,999\,999\,84 \\
&= +0.000\,000\,000\,037 \\
&> 0.
\end{aligned}$$

Therefore, the proposed quadratic equation has two distinct real roots.

19. Write the two-point equation of the line, and solve for v_0 where $y = 0$:

$$y - y_1 = \frac{y_2 - y_1}{x_2 - x_1}(x - x_1),$$

$$y - p_0 = \frac{q_0 - p_0}{w_0 - u_0}(x - u_0),$$

$$0 - p_0 = \frac{q_0 - p_0}{w_0 - u_0}(v_0 - u_0),$$

$$-p_0 * \frac{w_0 - u_0}{q_0 - p_0} = (v_0 - u_0),$$

$$u_0 - p_0 * \frac{w_0 - u_0}{q_0 - p_0} = v_0.$$

21. For instance, choose $u_0 = 1$ and $w_0 = 2$:

$$u_0 = 1, p_0 = f(u_0) = f(1) = 1^2 - 3 = -2 < 0 : (u_0, p_0) = (1, -2),$$
$$w_0 = 2, q_0 = f(w_0) = f(2) = 2^2 - 3 = 1 > 0 : (w_0, q_0) = (2, 1),$$
$$v_0 = u_0 - p_0 * \frac{w_0 - u_0}{q_0 - p_0} = 1 - (-2) * \frac{2 - 1}{1 - (-2)} = 1 + 2 * \frac{1}{3} = \frac{5}{3} = 1.666 \ldots.$$
$$f(v_0) = f(5/3) = (5/3)^2 - 3 = 25/9 - 3 = 25/9 - 27/9 = -2/9 < 0.$$

Thus, $u_1 = v_0 = 5/3$, and the new interval is $[u_1, w_1] = [5/3, 2]$. Repeat the same process:

$$u_1 = 5/3, p_1 = f(u_1) = f(5/3) = -2/9 < 0 : (u_1, p_1) = (5/3, -2/9),$$
$$w_1 = w_0 = 2, q_1 = f(w_1) = f(2) = 2^2 - 3 = 1 > 0 : (w_1, q_1) = (2, 1),$$
$$v_1 = u_1 - p_1 * \frac{w_1 - u_1}{q_1 - p_1} = 5/3 - (-2/9) * \frac{2 - 5/3}{1 - (-2/9)} = 5/3 + (2/9) * \frac{1/3}{11/9}$$
$$= 5/3 + 2/33 = \frac{57}{33} = 1.72\,72\,72 \ldots.$$
$$f(v_1) = f(57/33) = (57/33)^2 - 3 = -2/121 < 0.$$

Thus, $u_2 = v_1 = 57/33$, and the new interval is $[u_1, w_1] = [57/33, 2]$. Repeat the same process:

$$u_2 = 57/33, p_2 = f(u_2) = f(57/33) = -2/121 < 0 : (u_2, p_2) = (57/33, -2/121),$$
$$w_2 = w_1 = 2, q_2 = f(w_2) = f(2) = 2^2 - 3 = 1 > 0 : (w_2, q_2) = (2, 1),$$
$$v_2 = u_2 - p_2 * \frac{w_2 - u_2}{q_2 - p_2} = 57/33 - (-3/121) * \frac{2 - 57/33}{1 - (-2/121)} = 57/33 + (2/121) * \frac{9/33}{123/121}$$
$$= 57/33 + 6/1,353 = \frac{2,343}{1,353} = 1.731\,707\,317 \ldots.$$
$$f(v_2) = f(2,343/1,353) = (2,343/1,353)^2 - 3 < 0.$$

Thus, $u_3 = v_2 = 2,343/1,353$, $w_3 = w_2 = 2$ and the new interval is $[u_2, w_2] = [2,343/1,353, 2] = [1.731 \ldots, 2 \ldots]$.

23. See also **Figure 4**.

$$X^3 + 0.1000269X^2 - 8.070\,000\,100\,7 \times 10^{-7}X - 2.70883 \times 10^{-19} = 0.$$

At $X = U_0 = 0$ the left-hand side is negative, while at $X = W_0 = 10^{-5}$ it is positive. the method of false position leads to $X \approx 8.067 * 10^{-6}$, which corresponds to $\mathrm{pH} = -\log_{10}(8.067 * 10^{-6}) \approx 5.093\ldots$ as corroborated in MacLeod [1984]. See item (4) below for greater accuracy, starting from the following computational detail to locate the positive zero of

$$F(X) = X^3 + 0.1000269X^2 - 8.070\,000\,100\,7 \times 10^{-7}X - 2.70883 \times 10^{-19}.$$

1. Because the calculator produced the estimate $\tilde{X} \approx 8.067 * 10^{-6}$, it may be prudent to verify the nearest powers of ten, 10^{-6} and 10^{-5}, to verify that $10^{-6} < X < 10^{-5}$:

$$F(10^{-6}) = 10^{-18} + 0.1000269 * 10^{-12} - 8.070\,000\,100\,7 \times 10^{-7} * 10^{-6}$$
$$- 2.70883 \times 10^{-19}$$
$$< 10^{-18} + 1.0003 * 10^{-13} - 8.0 * 10^{-13} - 2.7 \times 10^{-19}$$
$$= 10 * 10^{-19} + 1.0003 * 10^{-13} - 8.0 * 10^{-13} - 2.7 \times 10^{-19}$$
$$= 7.3 * 10^{-19} - 7.0007 * 10^{-13} < -6.007 * 10^{-13}$$
$$< 0;$$
$$F(10^{-5}) = 10^{-15} + 0.1000269 * 10^{-10} - 8.070\,000\,100\,7 \times 10^{-7} * 10^{-5}$$
$$- 2.70883 \times 10^{-19}$$
$$> 10^{-15} + 1.0 * 10^{-11} - 10.0 * 10^{-12} - 2.7 \times 10^{-19}$$
$$= 10^{-15} + 0 - 2.7 \times 10^{-19}$$
$$> 0.$$

2. It may also be convenient and more accurate to factor out a power of ten. To this end, let $x := 10^6 X$, so that $1 < x < 10$, and factor out 10^{-13} from F, so that

$$F(X) = X^3 + 0.1000269X^2 - 8.070\,000\,100\,7 \times 10^{-7}X$$
$$- 2.70883 \times 10^{-19}$$
$$= (10^{-6}x)^3 + 0.1000269(10^{-6}x)^2 - 8.070\,000\,100\,7 \times 10^{-7}(10^{-6}x)$$
$$- 2.70883 \times 10^{-19}$$
$$= 10^{-13} * \left\{10^{-5} * x^3 + 1.000269 * x^2 - 8.070\,000\,100\,7 * x\right.$$
$$\left. - 883 \times 10^{-6}\right\}$$
$$= 10^{-13} * f(x),$$
$$f(x) := 10^{-5} * x^3 + 1.000269 * x^2 - 8.070\,000\,100\,7 * x$$
$$- 2.70883 \times 10^{-6}.$$

3. This confirms that x lies in the narrower range $8 < x < 9$, so that $8 * 10^{-6} < X < 9 * 10^{-6}$:

$$f(9) := 10^{-5} * 9^3 + 1.000269 * 9^2 - 8.070\,000\,100\,7 * 9$$
$$- 2.70883 \times 10^{-6}$$
$$= 0.00729 + 1.000269 * 81 - 8.070\,000\,100\,7 * 9$$
$$- 2.70883 \times 10^{-6}$$
$$> 0.00729 + 1.0 * 81 - 8.1 * 9 - 3.0 \times 10^{-6}$$
$$= 0.00729 + 81 - 72.9 - 3.0 \times 10^{-6}$$
$$> 81 - 72.9 = 8.2 > 0;$$
$$f(8) := 10^{-5} * 8^3 + 1.000269 * 8^2 - 8.070\,000\,100\,7 * 8$$
$$- 2.70883 \times 10^{-6}$$
$$= 0.00512 + 1.000269 * 64 - 8.070\,000\,100\,7 * 8$$
$$- 2.70883 \times 10^{-6}$$
$$< 0.00512 + 1.001 * 64 - 8.07 * 8$$
$$- 0 \times 10^{-6}$$
$$= 0.00512 + 64.064 - 64.56$$
$$< 0.00512 - 0.504 < -0.498 < 0.$$

4. From $u := 8$ and $w := 9$, *regula falsi* applied to f produces $v := 8.060\,163\,699\,51$ and $f(v) < 0$. The method stabilizes and stays at $v_9 := 8.067\,179\,571\,27$, so that $\tilde{X} = 8.067\,179\,571\,27 * 10^{-6}$.

5. Try $x := 8.1$ to verify that $8 < x < 8.1$, so that $8 * 10^{-6} < X < 8.1 * 10^{-6}$:

$$f(8.1) := 10^{-5} * (8.1)^3 + 1.000269 * (8.1)^2 - 8.070\,000\,100\,7 * (8.1)$$
$$- 2.70883 \times 10^{-6}$$
$$> 0.00524961 + 1.0 * 64.81 - 8.1 * 8.1 - 2.70883 \times 10^{-6}$$
$$= 0.00524961 + 64.81 - 64.81 - 2.70883 \times 10^{-6}$$
$$> 0.00524961 + 0 - 3.0 \times 10^{-6} = 0.005\,246\,61$$
$$> 0.$$

6. Thus, $X = 8.05 * 10^{-6} \pm 0.05 * 10^{-6}$, with a relative error less than 1%:

$$\frac{|X - \tilde{X}|}{|X|} < \frac{0.05 * 10^{-6}}{8 * 10^{-6}} = 0.00625 = 0.625\%.$$

(The number of digits listed has nothing to do with accuracy: anyone who insists on writing either 8.0 or 8.1 doubles the error.)

References

Benjamin, Arthur. 1987. The bisection method. Which root? *American Mathematical Monthly* 94 (9): 861–863.

Birkhoff, Garrett, and Saunders Mac Lane. 1977. *A Survey of Modern Algebra.* 4th ed. 50th anniversary 1991 printing. New York: Macmillan.

Dedekind, Richard. 1963. *Essays on the Theory of Numbers.* New York: Dover. 1963. (Republication of the English translation from the German published by the Open Court Publishing Company in 1901.)

Edwards, Charles Henry. 1979. *The Historical Development of the Calculus.* New York: Springer-Verlag.

Etgen, Garret, reviser. 1999. *Salas and Hille's Calculus: One and Several Variables.* 8th ed. New York: Wiley.

Hungerford, Thomas W., and Richard Mercer. 1985. *Precalculus Mathematics.* 2nd ed. Philadelphia, PA: Saunders College Publishing.

Lang, Serge. 1965. *Algebra.* Reading, MA: Addison-Wesley.

Mac Lane, Saunders. 1997. On the Harvard Consortium Calculus. *Notices of the American Mathematical Society* 44 (8): 893.

MacLeod, Allan J. 1984. The calculation of pH. *International Journal of Mathematics Education in Science and Technology* 15 (6): 691–696.

Neugebauer, O. 1969. *Vorlesungen über die Geschichte der antiken mathematischen Wissenschaften. Erster Band: Vorgriechische Mathematik.* 2nd ed. Berlin: Springer-Verlag.

Nievergelt, Yves. 1995. Bisection hardly ever converges linearly. *Numerische Mathematik* 70 (1995): 111–118. Reviewed in *Mathematical Reviews* #95k:65055.

Ostebee, Arnold, and Paul Zorn. 1997. *Calculus from Graphical, Numerical, and Symbolic Points of View.* 2 vols. Fort Worth, TX: Saunders College Publishing.

————. 1998. *Multivariable Calculus from Graphical, Numerical, and Symbolic Points of View.* Fort Worth, TX: Saunders College Publishing.

Pan, Victor Y. 1997. Solving a polynomial equation: Some history and recent progress. *SIAM Review* 39 (2): 187–220.

Pulskamp, Richard J., and James A. Delaney. 1991. *Computer and Calculator Computation of Elementary Functions.* UMAP Modules in Undergraduate Mathematics and Its Applications: Module 708. Lexington, MA: COMAP, 1991. Reprinted in *The UMAP Journal* 12 (1991): 315–348. Reprinted in *UMAP Modules: Tools for Teaching 1991,* edited by Paul J. Campbell, 1–34. Lexington, MA: COMAP, 1992.

Schelin, C.W. 1983. Calculator function approximation. *American Mathematical Monthly* 90 (5): 317–325.

Michael. 1980. *Calculus*. 2nd ed. Wilmington, DE: Publish or Perish.

t, James. 1995. *Calculus*. 3rd ed. Pacific Grove, CA: Brooks/Cole.

Strang, Gilbert. 1991. *Calculus*. Wellesley, MA: Wellesley-Cambridge Press.

Wall Street Journal. 1993. Western Ed. 79 (25 August 1993): C21.

van der Waerden, Bartel Leenert. 1983. *Geometry and Algebra in Ancient Civilizations*. New York: Springer-Verlag.

About the Author

Yves Nievergelt graduated in mathematics from the École Polytechnique Fédérale de Lausanne (Switzerland) in 1976, with concentrations in functional and numerical analysis of PDEs. He obtained a Ph.D. from the University of Washington in 1984, with a dissertation in several complex variables under the guidance of James R. King. He now teaches complex and numerical analysis at Eastern Washington University.

Prof. Nievergelt is an associate editor of *The UMAP Journal*. He is the author of several UMAP Modules, a bibliography of case studies of applications of lower-division mathematics (*The UMAP Journal* 6 (2) (1985): 37–56), and *Mathematics in Business Administration* (Irwin, 1989). His new book is *Wavelets Made Easy* (Birkhäuser, 1999).